A C S   S Y M P O S I U M   S E R I E S **541**

Mass Spectrometry for the Characterization of Microorganisms

Catherine Fenselau, EDITOR

University of Maryland Baltimore County

Developed from a symposium sponsored
by the Division of Analytical Chemistry
at the 204th National Meeting
of the American Chemical Society,
Washington, DC,
August 23–28, 1992

American Chemical Society, Washington, DC 1994

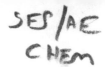

Library of Congress Cataloging-in-Publication Data

Mass spectrometry for the characterization of microorganisms:
developed from a symposium sponsored by the Division of Analytical
Chemistry at the 204th National Meeting of the American Chemical
Society, Washington, DC, August 23–28, 1992 / Catherine Fenselau,
editor.

    p.   cm.—(ACS symposium series, ISSN 0097–6156; 541)

    Includes bibliographical references and index.

    ISBN 0–8412–2737–3

    1. Mass spectrometry—Congresses. 2. Microorganisms—
Identification—Congresses. 3. Microbiological chemistry—
Congresses.

    I. Fenselau, Catherine, 1939–   . II. American Chemical Society.
Division of Analytical Chemistry. III. American Chemical Society.
Meeting (204th: 1992: Washington, D.C.) IV. Series.

QR69.M33M37   1994
576′.072—dc20
                                            93–39043
                                               CIP

The paper used in this publication meets the minimum requirements of American National
Standard for Information Sciences—Permanence of Paper for Printed Library Materials, ANSI
Z39.48–1984. ∞

PRINTED IN THE UNITED STATES OF AMERICA

Foreword

THE ACS SYMPOSIUM SERIES was first published in 1974 to provide a mechanism for publishing symposia quickly in book form. The purpose of this series is to publish comprehensive books developed from symposia, which are usually "snapshots in time" of the current research being done on a topic, plus some review material on the topic. For this reason, it is necessary that the papers be published as quickly as possible.

Before a symposium-based book is put under contract, the proposed table of contents is reviewed for appropriateness to the topic and for comprehensiveness of the collection. Some papers are excluded at this point, and others are added to round out the scope of the volume. In addition, a draft of each paper is peer-reviewed prior to final acceptance or rejection. This anonymous review process is supervised by the organizer(s) of the symposium, who become the editor(s) of the book. The authors then revise their papers according to the recommendations of both the reviewers and the editors, prepare camera-ready copy, and submit the final papers to the editors, who check that all necessary revisions have been made.

As a rule, only original research papers and original review papers are included in the volumes. Verbatim reproductions of previously published papers are not accepted.

M. Joan Comstock
Series Editor

Contents

Preface

AMONG PHYSICOCHEMICAL TECHNIQUES, mass spectrometry offers the unique combination of great speed, high specificity, and excellent sensitivity. These features are all required for detection and characterization of microorganisms when the public health is threatened by contaminated food or polluted air and water, in hospitals caring for increasing numbers of immunodeficient patients, and in battlefield situations in the New World Order. Rapid and reliable analysis of microorganisms is also useful in the petroleum, biotechnology, and other industries.

This book provides specific discussions of the classes of biomarkers to which mass spectrometry has been applied, including nucleosides, proteins, carbohydrates, phospholipids, glycopeptidolipids, and lipopolysaccharides from bacteria, viruses, and fungi. Ongoing major contributions by mass spectrometry are also related to the structure elucidation of potential new chemical markers. In most cases, newly determined structures are being correlated with biochemical and pathological functions. Chemical markers for antigenic activity, for example, might be more useful in some situations than phylogenetic markers.

Capabilities and appropriate roles are illustrated for gas chromatography–mass spectrometry, high-pressure liquid chromatography–mass spectrometry, tandem mass spectrometry, desorption mass spectrometry, and pyrolysis mass spectrometry. Some evolving technologies are previewed. Opportunities exist for development of specialized or dedicated instruments, especially with automated systems for preparation and introduction of samples.

This volume and the symposium from which it was developed are intended to assess the capabilities emerging for identification of microorganisms by mass spectrometry, to discuss the kinds of characterizations required in different situations, and to project directions for future growth.

Acknowledgments

I express my special appreciation to David White, who worked with his co-authors to produce a chapter from a hospital bed following serious injury in an automobile accident. In addition to leading the field of chemotaxonomy of microorganisms, David is a dedicated professional scientist and a good friend. I also want to acknowledge and thank the scientists who helped to review the manuscripts, all active contributors to this field in their own right: Jaap Boon, Duncan Bryant, Joan Bursey, Elaine Fukuda, Beth Gillece-Castro, David Heller, Sanford Markey, Steve Morgan, Karl Schram, Philip B. Smith, Anthony Tsarbopoulos, and Sue Weintraub. All the authors are indebted to Sue McCain for her cheerful assistance at University of Maryland Baltimore County and to Rhonda Bitterli for her highly professional support at the American Chemical Society.

CATHERINE FENSELAU
Department of Chemistry and Biochemistry
University of Maryland Baltimore County
Baltimore, MD 21228

October 7, 1993

Chapter 1

Mass Spectrometry for Characterization of Microorganisms

An Overview

Catherine Fenselau

Department of Chemistry and Biochemistry, University of Maryland Baltimore County, 5401 Wilkens Avenue, Baltimore, MD 21228

The advantages and disadvantages of mass spectrometry for the characterization of microorganisms are summarized. Different families of chemotaxonomic markers are reviewed, especially with respect to mass spectrometric analysis. An overview is provided to the chapters in the book.

The identification of microorganisms has been an important aspect of public health since the connection was first recognised between bacteria and disease. Characterization has also become important for such diverse objectives as the detection of biological warfare agents, protection of stored crops, monitoring of bioprocessors, and evaluation of potential oil drilling sites. Often the analytical objective is simply to detect the presence of microorganisms and to distinguish between bacteria, viruses or fungi. In other cases, species or even strains must be identified. Historically such identifications have depended heavily on microscopic examination of homogeneous samples of the microorganisms obtained by culturing under laboratory conditions. Mixed populations are problematic, in part because a species of lessor interest might be more readily cultured. In recent years, automated instruments have been commercialized to characterize microorganisms on the basis of a battery of enzymatic reactions monitored spectroscopically. A chemotaxonomic characterization has also been automated and commercialized for bacteria, based on analysis of fatty acid content by gas chromatography. Both of these approaches require culturing homogeneous populations under standard conditions through several days. Gene characterization based on the use of DNA probes or gene amplification by the polymerase chain reaction can be used to test for predicted species in a matter of hours. These techniques for gene characterization cannot search for a large set of possible identities in a single analysis.

An ideal analytical method would be one that could be applied to a broad set of microorganisms and to mixed populations in a few minutes. Mass spectrometry analyzes chemical markers for a broad set of organisms and from

0097–6156/94/0541–0001$06.00/0

mixed populations. Only minutes are required for the instrumental measurement, however, the amount of culturing and sample preparation required varies with the situation, as is discussed throughout this book. It is this author's opinion that the various approaches are, in fact, complementary, and that the best strategies will integrate chemotaxonomy, morphology and gene probes. Particularly in forensic analysis,for example, chemotaxonomic analyses might rapidly place the microorganism within a limited group, such as gram negative bacteria, and direct the efficient use of DNA probes for more precise identification.

Mass Spectrometry Techniques

Further development of methods for instrument-based chemotaxonomic identification of microorganisms is underway around the world. Automated physicochemical instrumentation offers high reliability, rapid throughput and increasing capability to analyse mixed populations. Among these techniques, the various configurations of mass spectrometry offer the combined advantages of speed, sensitivity, specificity and selectivity. Its cost and complexity have historically been considered disadvantagous for widespread use in hospitals or mobile environmental units.

Mass spectrometry was first evaluated in the 1970's as a technique to fingerprint microorganisms. At that time the available instrumentation required that molecules be vaporized into the gas phase before they could be ionized and analyzed. In one early study (1) controlled heating of dehydrated bacteria was used to release molecules in the mass range above 700 atomic mass units. Other laboratories introduced high temperature pyrolysis of intact bacteria, which converts chemical markers to gases with molecular weights below 150 atomic mass units (2).

The development of desorption techniques for mass spectrometry, plasma desorption (3,4) laser desorption (5,6) and fast atom bombardment (7,8), allowed involatile samples to be ionized and brought into the gas phase. When these techniques were applied to intact or solvent-lysed bacteria, phospholipids and other polar lipids were found to be preferentially desorbed and analysed (9,10). Interest in the characterization of phospholipids in bacteria led to the first consideration of the problems associated with compiling reference libraries of fast atom bombardment spectra (11), an example of how a challenging anaytical problem can stimulate methods development.

Tandem mass spectrometry has been coupled very effectively with desorption techniques for analyses of intact or lysed microorganisms. Such instrumentation permits separation of mixtures on the basis of masses, and significant improvement of signal to noise ratios. Activation of selected precursor ions in a cell between the two mass analysers provides more extensive and reproduceable fragmentation (12-14). This activation is usually carried out by collisions with inert gases, however, reactive collisions and photoexcitation with lasers are under development. Recognition of polar lipid classes can be significantly facilitated by scanning tandem instruments to detect precursor ions of a characteristic fragment ion, or to record precursor ions that lose a characteristic neutral moiety (constant neutral loss scans) (15).

Other kinds of chemical markers can also be characterized by mass spectrometric techniques, e.g., peptidoglycans (16), carbohydrates (chapter 13 this volume), glycolipids (chapters 12, 14), and proteins (chapter 11). Usually these must first be isolated from the microorganism. Recently, electrospray ionization (17,18) and matrix assisted laser desorption (19) have provided molecular weight determinations above 100,000 daltons by mass spectrometry, and the incorporation of accurate mass balance strategies into biopolymer sequence determinations. These techniques have been applied to proteins and carbohydrates isolated from viruses, bacteria and other microorganisms.

Some of the most powerful applications of mass spectrometry involve mass spectrometry interfaced with a separation technique, most commonly gas chromatography or high pressure liquid chromatography (20, 23). Such a combined instrument provides increased capability for mixtures, as well as two complementary kinds of characterization. Chromatographic separations have the important ability to distinguish chirality, to which mass spectrometry is usually blind.

Chemotaxonomic Markers

Among the endogenous molecules in a microorganism, deoxynucleic acid (DNA) is usually considered to be the most individually characteristic class. Developments are underway of antibody and other assays based on amplified DNA to permit recognition of familiar species, even in mixed populations. Although mass spectrometric methods have not to this time been well developed for gene sequencing, its capabilities have been applied with striking results to the elucidation of structures of novel nucleosides in nucleic acids. The potential of nuclease digests of tRNA as phylogenetic markers is discussed elsewhere in this book (chapter 10) by the research group that has developed this area of structure study. Combined high pressure liquid chromatography-mass spectrometry instrumentation has been critical in these structure studies, as well as the capability to form molecular ions from involatile samples and to obtain fragmentation reproduceably.

Protein structures can also serve to distinguish closely related microorganisms, and the use of protein sequences to characterize viral variants that replicate successfully in rapidly mutating populations is illustrated in this book (chapter 11). In addition to molecular weight determinations of proteins by matrix assisted laser desorption or electrospray, peptide products of protease digests can be mapped by their molecular weights, and peptides can be sequenced by collisional activation in tandem mass spectrometry experiments. In situations where a gene sequence has already been determined, its speed makes mass spectrometry the method of choice to confirm translation and identify processing and post-translational modifications by protein mapping and sequencing.

The complex polar lipids constitute a third class of molecules with chemotaxonomic utility and the only class of biomarkers directly accessible in intact microorganisms by mass spectrometry. Free fatty acids and fatty acid components of complex polar lipids vary with health, life cycle and nutrition, however the polar head groups of phospholipids are qualitatively and quantitatively stable enough to be considered taxonomically characteristic. Polar lipids are abundant, constituting as much as 5% of bacterial dry weight, for example, and their amphiphillic properties make them easy to recover and to analyse by desorption mass spectrometry. For these and other reasons mass spectrometric methods for analysis of biomarkers of microorganisms have been widely developed for phospholipids and other polar lipids. The scheme shows a simple procedure developed in this laboratory (21) to screen different kinds of microorganisms, based on the predominant phospho- or other polar lipid detected in their fast atom bombardment mass spectra.

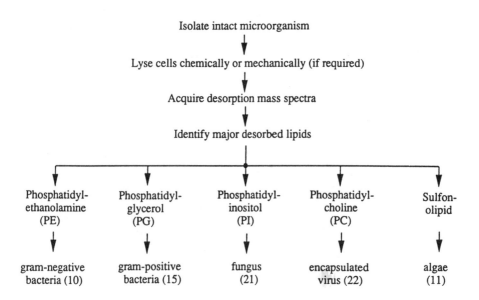

Characterization requires that the distinctive polar head groups be recognised as part of complex lipid molecules, yet characterized independent of the more variable fatty acid moieties. Since the sets of structural variables (fatty acids, polar head groups) are usually bounded, assignments can be made on the basis of molecular weights deduced from molecular ions (which are abundant). Polar lipids also undergo fragmentation in which the characteristic polar head group is eliminated. This pattern allows superior molecular characterization of each family of polar lipids by tandem mass spectrometry experiments using collisional activation and constant neutral loss scans (15). Detailed qualitative and quantitative analyses of polar head groups can provide finer distinctions, e.g., of genus or species, and computer-supported mathematical processing of spectra has the potential to identify mixed populations (11, 24).

Analysis of fatty acids within the complex lipids may provide a view of the nutritional and metabolic history of the microorganism. For example, cultured human immunodeficiency virus has been shown by tandem mass spectrometry experiments to contain phyosphatidylcholine with fatty acid distributions similar to those of their mammalian cell hosts (22). More than a dozen groups have contributed to the development of chemotaxonomy based on mass spectrometric analysis of polar lipids, and three of these share their experience in this book (chapters 2,3,4).

Gram negative bacteria carry on their cell surfaces large quantities of complex glycolipids (called lipooligosaccharides or lipopolysaccharides, depending on their size) which contain the major antigenic sites and control other pathogenic interactions in mammalian hosts. Some mycobacteria carry glycopeptidolipids and phenolic glycolipids on their surfaces, which also evoke specific antibodies and provide the basis for serodiagnosis. Thus knowledge of the structures of these glycolipids has important implications for controlling diseases such as gonorrhea and tuberculosis, as well as for chemotaxanomic differentiation. All of the techniques of mass spectrometry have been used in the difficult structure studies of these complex molecules and their carbohydrate and lipid A components isolated from various kinds of cultured bacteria. The analytical targets, effective techniques and biomedical implications are discussed in three chapters here (chapters 12,13,14).

Another approach that exploits the unique chemical compositions of microbial cell walls and membranes involves recognition of chemical markers in whole sample hydrolysates. Gas chromatography-mass spectrometry and high pressure liquid chromatography-mass spectrometry are considered to provide the greatest potential for this kind of mixture analysis. For example, muraminic acid and d-alanine are present in high concentrations in the crosslinked peptidoglycan structures of most eubacteria, but rarely encountered in plants or animals. Unique monosacharides that can be exploited as chemical markers for bacteria, are described by Fox (chapter 8). The potential of such approaches has been recognized for monitoring microbial contaminations in the fermentation industry (25) and for noninvasive clinical diagnosis (26). Recently, a rapid, specific and sensitive method for quantitation of candidiasis in hospital patients has been developed and applied by Roboz. Its basis, the gas chromatography-mass spectrometric quantitation of ratios of d-and l-arabinitol, and its validation are discussed elsewhere in the book (chapter 9).

Some approaches have received serious consideration because of their simplicity rather than their chemotaxonomic strength. Reports on the development of headspace analysis (chapter 6) and membrane diffusion (chapter 7) to detect microorganisms are included in the book, as well as a report from a U.S. Army laboratory on the status of pyrolysis analy sis (chapter 5). This author agrees with the observations of Gutteridge (2) and others that the use of pyrolysis to convert intricate endogenous chemical markers to small gaseous molecules is philosophically flawed and has been historically unproductive.

Mass Spectrometry in Various Strategies

As this volume illustrates, there are three general levels at which analysis of microorganisms is usually carried out. Stategies for screening for the presence of contamination or infection, usually with the expectation to discriminate yeast from bacteria from viruses, often are based on fingerprinting small metabolic markers in air or solution. The specificity, selectivity, speed and sensitivity of mass spectrometry combined with chromatography will continue to make it the method of choice for this kind of trace analysis. Any large scale application, e.g., hospital fungal infections, will stimulate development of customized and simplified instrumental systems.

A second analytical level involves identification of microorganisms often including their genus and species. Here a variety of morphologic and chemotaxonomic techniques are in use, most of which require culturing the microorganisms as a way to achieve bulk separation from a matrix such as soil or tissue, to isolate pure samples of each species or strain, or to increase the amount of material available for study. The requirement for culturing in chemotaxonomic strategies based on mass spectrometry depends on the level of contamination. Bacteria would need to be cultured out of hamburger, for example, but might be identified directly as filtered from the atmosphere. In contrast to chemotaxonomic approaches that use DNA probes, reference libraries of mass spectra can provide identification of many potential candidates from a single analysis.

The third class of analysis is structure elucidation of unknown molecules in fully characterized microorganisms. Mass spectrometry will continue to be a lead technique in structure determination and structure/function studies of complex biomolecules.

Literature Cited

1. Anhalt, J.P., Fenselau, C. *Anal. Chem.* **1975**, 47, 219.
2. Gutteridge, C.S. *Methods in Microbiology* **1987**, 19, 227.
3. Macfarlene, R.O., Skowronski, R.P., Torgerson, D.F., *Biochem. Biophys. Res. Commun.* **1974** 60, 616.
4. Cotter, R.J. *Anal. Chem.* **1988**, 60, 781A.
5. Barber, M., Bordoli, R.S., Sedgwic, R.D., Tyler, A.N. *J. Chem. Soc. Chem. Commun.* **1981**, 325.
6. Fenselau, C., Cotter, R.J., *Chem. Rev.* **1987**, 57, 501.
7. Posthumus, M.A., Kistemaker, P.G., Meuzelaar, H.L.C., *Anal. Chem.* **1978**, 50,985.
8. Cotter, R.J. *Analytica Chimica Acta.* **1987**, 195, 45.
9. Heller, D.N., Fenselau, C., Cotter, R.J., Demirev, P., Olthoff, J.K., Honovich, J., Uy, M. Tanaka, T., Kishimoto, Y. *Biochem. Biophys. Res. Commun.* **1987**, 142, 194.

10. Heller, D.N., Cotter, R.J., Fenselau, C., Uy, O.M. *Anal. Chem.* **1987**, 59, 2806.

11. Platt, J., Uy, O.M., Heller, D.N., Cotter, R.J., Fenselau, C. *Anal. Chem.* **1988**, 60, 1415.

12.*Tandem Mass Spectrometry;* McLafferty, F.W. Eds.; John Wiley & Sons, NY, 1983.

13. *Mass Spectrometry/ Mass Spectrometry: Techniques and Applications of Tandem Mass Spectrometry;* Busch, K.L., Glish, G.L., McLuckey, S.A., Eds.; VCH Publishers, Inc.: NY, NY, 1988.

14. Fenselau C. In *Annual Rev. Pharmacol. and Toxicol.*; Cho, A., Blaschke, T, Loh, H. Way, J., Eds.; Annual Reviews Inc.: Palo Alto, CA, 1992; 32, 555-578.

15. Heller, D.N., Murphy, C.M., Cotter, R.J., Fenselau, C, Uy, D.M. *Anal. Chem.* **1988**, 60, 2787.

16. Martin, S.A., Rosenthal, R.S., Biemann, K. *J. Biol. Chem.* **1987**, 262, 7514.

17. Fenn, J.B., Mann, M., Meng, C.K., Wong, S.F., Whitehouse, C.M., *Science* **1989**, 246, 64.

18. Loo, J.A., Edmonds, C.G., Smith, R.D., *Science*, **1990**, 248, 201.

19. Hillenkamp, F, Karas, M., Beavis, R.C., Chait, B.T. , *Anal. Chem.* **1991**, 63, 1193A.

20. Odham, G., Valeur, A., Michelsen, P. Aronsson, E., McDowall, M., *J. Chromatog.* **1988**, 434, 31.

21. Heller, D.N., Murphy, C.M., Cotter, R.J., Platt, J.A., Uy, O.M., Fenselau, C. In *Advances in Mass Spectrometry*; Longevialle, P. Ed.; Heydon & Sons Ltd.: London, 1989, 11B, 1612.

22. Bryant, D.K, Orlando, R.C., Fenselau, C., Sowder, R.C., Henderson, L.E., *Anal. Chem.* **1991**, 63, 1110.

23. Poon, G.K., Griggs, L.J., Edwards, C., Beattie, K.A., Codd, G.A. *J. Chromatography* **1993**, 628, 1215.

24. Cole, M.J., Hemenway, E.C., Enke, C. G., Paper presented at the 39th Annual Conference on Mass Spectrometry and Allied Topics, Nashville, TN, May 16-24, 1991; Cole, M.J., Enke, C.G. Chapter 3 this volume.

25. Elmroth, I., Valeur, A., Odham, G., Larsson, L. *Biotechnology and Bioengineering* **1990**, 35, 787.

26. French, G.L., Chan, C.Y, Poon, D., Cheng, S.W., Cheng, A.F. *J. Med. Microbiol.* **1990**, 31, 21.

RECEIVED September 20, 1993

Chapter 2

Rapid Identification of Microbes from Clinical and Environmental Matrices

Characterization of Signature Lipids

D. C. White[1,2], D. B. Ringelberg[1], D. B. Hedrick[1], and D. E. Nivens[1]

[1]Center for Environmental Biotechnology, University of Tennessee, Knoxville, TN 37932–2567
[2]Environmental Sciences Division, Oak Ridge National Laboratory, Oak Ridge, TN 37831, and Department of Microbiology, University of Tennessee, Knoxville, TN 37996–0845

Lipid biomarkers can be recovered from isolates, clinical specimens, and environmental samples by a single-phase chloroform/methanol extraction, fractionation of the lipids on silicic acid, and derivatization prior to analysis by capillary gas-liquid chromatography/mass spectrometry. Polar lipid fatty acid profiles are a quantitative measure of bacterial and eukaryotic viable biomass and community structure, while sterols are used to measure the eukaryotes and isoprenoid ether lipids the archaea. Changes in a microbe's environment are reflected in the adaptive responses of its lipids, and can be used to delineate details of the microbe's metabolic status and environmental conditions. Rapid methods of sample preparation such as supercritical fluid extraction will extend the range of application of lipid biomarker analysis.

Since every cell has a lipid membrane, lipid composition varies with species, and cellular lipid modification is necessary to adapt to environmental conditions, microbial lipids are indicative of microbial biomass, community structure, and metabolic status. Rapid methods are being developed which will expand the range of applicability of lipid biomarker analysis by reducing time, expense, and waste production per sample analyzed, as well as (in some applications) increasing analytical sensitivity (1). Mass spectroscopy (MS) is the analytical detection system of choice for many of these applications due to its high sensitivity and selectivity.

In this review, specific classes of lipid biomarkers and examples of their application, approaches to their rapid analysis, and some areas of potential application will be described. Mass spectral analysis of intact phospholipids, glycolipids, and lipopolysaccharides, as well as pyrolysis MS are considered in other chapters of this symposium.

0097–6156/94/0541–0008$06.00/0

Signature Lipid Biomarkers

A biomarker indicates the presence or metabolic activity of biota in an environment or sample. Molecular biology (2-5), immunology (6), and a variety of spectroscopies (7, 8) have been used to detect biomarkers. Geochemists determine the sources and diagenic history of organic matter in sediments by analysis of hydrocarbon biomarkers (9). Microbial lipid biomarkers have been established for bacterial, archaeal, and eukaryotic viable biomass, and for determining the microbial community structure of a sample. Since microbial lipids are phenotypic characters, they respond to environmental conditions and can be used to measure metabolic status. Lipid biomarkers are easily extracted from sample matrix materials (10), and may be readily introduced to a mass spectrometer by gas chromatography (GC) (11) or supercritical fluid chromatography (12, 13).

Measures of Microbial Biomass. The most basic measure of a microbial community is the total microbial biomass, but the traditional microbiological methods of bacterial enumeration by most probable number, plate counts, and direct counts are limited in their applicability for rapid analysis. Most probable number and plate count methods require culturing of the organisms, which under-report the biomass due to the inability of any media to support all organisms (14), and the presence of bacteria which are metabolically active but not culturable (15). Direct count of microbes by microscopy is very difficult or impossible on sample matrices with high solids content (16), and generally suffer high variability (17, 18).

Phospholipids are a reliable measure of the viable biomass of a sample for three reasons: they are components of every cell membrane but are not metabolic storage products (19), they are readily separated from other natural or anthropogenic compounds by their amphipathic nature (20), and are rapidly degraded upon cell death (21). Bacterial and eukaryotic polar lipids contain ester-linked fatty acids, while the archaea (formerly the archaebacteria, 22) use exclusively ether-linked lipids in their membranes. Ester- and ether-linked polar lipids will be dealt with separately.

Polar Lipid Fatty Acids. A comparison of polar lipid fatty acids (PLFA) with adenosine triphosphate as biomarkers for viable microbial biomass in an uncontaminated subsurface aquifer sediment (23) both gave essentially the same value for cell numbers as direct counts, using conversion factors derived from cultures of subsurface bacteria. However, the standard deviations of the biomass estimates by direct counts and adenosine triphosphate were 17 and 5 times higher than for PLFA, respectively.

Ether Lipids. One of the characteristic features of the archaea is the presence of ether-linked membrane lipids (24), rather than the ester-linked membrane lipids utilized by bacteria and eukaryotes. The high molecular weight of the archaeal diethers and tetraethers makes GC impractical, but they are readily analyzed by supercritical fluid chromatography (25, 26). A sample preparation protocol was developed which allowed the simultaneous determination of archaeal ether lipids as well as bacterial and eukaryotic PLFA in the same sample, conserving valuable sample material. This approach also increased the power of statistical analysis by providing PLFA and ether lipid data on the same sample rather than on parallel samples. The combination of PLFA and archaeal ether lipid analysis has been applied to methanogenic bioreactors (27, 28) and deep-sea hydrothermal vent ecosystems (29), and fresh-water sediment communities (30). Other archaeal ether lipid analyses include methanogens in submarine hypersaline basins (31) and in fresh water sediments (32)

Using this method (29), a transect of 4 samples through the metal sulfides of a deep-sea hydrothermal vent spanning from the 2°C ambient seawater to the ~350°C hydrothermal vent fluids was analyzed. The highest bacterial biomass was detected in the sample adjacent to the 2°C seawater where archaea were undetectable. Archaeal biomass was higher than bacterial in the next sample of the transect, comparable to bacterial biomass in the seawater-exposed sample. Both measures of biomass decreased greatly in the next sample, becoming nearly undetectable in the hydrothermal vent fluid-adjacent sample.

Sterols. Microeukaryotes have polar lipids that are generally not as structurally diverse as bacteria, so their PLFA are not as useful for identification. Fungi and protozoa can be classified by their membrane sterols (33). The use of sterols as a measure of microeukaryote biomass has been tested using estuarine sediment in the laboratory (34). In one treatment eukaryotic growth was stimulated by addition of sucrose and nutrient broth and bacterial growth was inhibited by the addition of penicillin and streptomycin. In the second treatment bacterial growth was stimulated by phosphate and glutamine while and eukaryotic growth was inhibited with cycloheximide. The ratio of sterols to lipid phosphate was much greater in the first treatment than the second, as predicted by the physiological requirements of fungi versus bacteria, and as shown by microscopic examination.

Measures of Microbial Community Structure. A microbial community's structure is the proportions its component viable organism populations. The pattern of individual fatty acids in the PLFA profile is representative of the microbial community contributing to the profile. The ability to identify the position and conformation of the double bond in monoenoic PLFA has greatly increased the specificity of the signature biomarker analytical system (35). Eukaryotes predominantly synthesize $\omega 9$ cis monounsaturated PLFA (where $\omega 9$ indicates the unsaturation is 9 carbons from the methyl end of the fatty acid). Bacteria most often make the $\omega 7$ isomers. Bacteria also form monoenoic PLFA with other positions of monounsaturation such as $\omega 5$ and $\omega 8$, which are characteristic of specific groups of microbes (36-38).

Identification of Specific Organisms. Bacterial species are routinely identified in pure culture by their PLFA (1). In certain clinical samples (blood, urine, or cerebrospinal fluid) any bacterium is a suspect pathogen. Larsson et al. (39) have successfully utilized the detection of the PLFA tuberculostearic acid (10-methyloctadecanoate) in sputum for the diagnosis of miliary tuberculosis. The pentafluorobenzyl ester of tuberculosteric acid was formed after saponifying the sputum, and was detected by GC-negative ion chemical ionization-MS (methane). The correct 7 sputum samples out of 23 coded samples were determined to contain *Mycobacterium*. This is a significant advance given the unreliability of direct microscopic examination of sputum smears, and the weeks which may be required to cultivate this slow-growing organism (*39*).

Community Structure by Fingerprint Analysis. The complexity of environmental microbial consortia makes the identification of every species in a sample impossible. Changes in the pattern of PLFA in samples with different treatments can be used to establish changes in community structure without the necessity of determining the species making up the community. Whether identification of species or fingerprint analysis of PLFA is used depends upon the nature of the sample analyzed and questions to be answered.

An example of the fingerprint approach to community structure analysis by PLFA patterns was a recent study of the microbial community in the lining of the burrow of the estuarine crustacean *Callianassa trilobata*, compared with unmodified ambient sediment (*40*). The burrow lining relative to the ambient sediment was enriched in total biomass, and relatively enriched in aerobic heterotrophs and sulfate reducing bacteria.

Metabolic Status of Microbial Consortia. The metabolic status of microbiota is a result of environmental conditions such as nutrient availability, temperature, and pH. Microbial adaptations to the environment are reflected in their lipid profiles. The ability to determine the metabolic status of microbial consortia greatly increases the subtlety of the information available from signature lipid biomarker analysis.

Respiratory Quinones. A class of neutral lipids with great potential as signature lipid biomarkers are the respiratory quinones (*41, 42*). The type of quinone elaborated by a bacterium depends upon the electron acceptor used in respiration (oxygen, sulfate, ferrous iron) while bacteria fixing energy by fermentation to not synthesize respiratory quinones (*43*). Respiratory quinones have been analyzed by liquid chromatography with electrochemical (*43*) and UV-visible absorption (*44*). While they are too large and thermally labile for GC, they are readily separated by supercritical fluid chromatography (*45*).

Cyclopropyl and Trans Mono-unsaturated Fatty Acids. Bacteria respond to starvation by synthesizing unique PLFA with cyclopropyl (*46*) or *trans*-unsaturated (*47*) moieties. These valuable biomarkers are detected as part of the PLFA profile by GC-MS.

Poly-ß-hydroxyalkanoates. Bacteria accumulate the polymer poly-ß-hydroxyalkanoates under conditions of unbalanced growth, that is, when there is plentiful carbon and energy substrates but cell division is prevented by the lack of a required nutrient such as nitrogen or phosphorous (*48*). In marine sediments, mechanical disturbance or the addition of natural chelators increased the ratio of poly-ß-hydroxyalkanoates to PLFA (*49*). Mechanical disturbance exposed the microbiota to fresh carbon and energy substrates leading to the exhaustion of one or more nutrients required for cell division. Chelators increased the ratio of poly-ß-hydroxyalkanoates to PLFA by making required trace metals unavailable to bacteria, inhibiting cell division.

Triglycerides. Neutral lipid triglycerides are accumulated as carbon and energy storage compounds by eukaryotes. There is evidence that in some species, triglycerides are accumulated preparatory to cell division (*50*). The triglyceride content of samples is quantified as the neutral lipid glycerol or ester-linked fatty acids by GC (*51*), or as intact triglycerides by supercritical fluid chromatography (*52*) or high temperature GC (*53*).

Rapid Analysis of Lipid Biomarkers

MIDI System. One approach to the rapid analysis of bacterial lipids is the commercially available MIDI system (*54*). Under standardized conditions, bacteria are cultured and harvested, and their whole cell fatty acids are quantified by saponification, methylation, and GC-MS with an autosampling system. Since the culturing step requires 24 to 48 hours, and the procedure requires a pure culture to begin with, the time required to obtain an identification is from two to four days. The system comes with a library of bacterial fatty acid profiles and the software to calculate similarities between profiles and determine identifications. This system has been used in clinical applications (*55*), identification of plant diseases (*56*), and microbial ecology (*57*).

Supercritical Fluid Extraction. Supercritical fluid extraction (SFE) can replace lipid extraction procedures using chloroform/methanol (*10*) in lipid biomarker analysis. Increasing the speed, reducing the expense, and eliminating most of the waste solvents would make lipid biomarker analysis more attractive in many areas, and could open up totally new applications.

SFE was used to replace soxhlet extraction of free fatty acids, sterols, sterol esters, and triglycerides from hamster feces in a nutritional study (*13*). Analysis of extracts was by supercritical fluid chromatography. Where soxhlet extraction required 3 days and 500 mL of solvent per sample, SFE required 1 hour and 2 mL of solvent. All components recovered by soxhlet extraction were found in SFE extracts, at from 9.2% to 17.2% higher levels, depending upon the SFE conditions used.

Derivatization/Supercritical Fluid Extraction. The membrane lipids found in the glycolipid and polar lipid fractions are not soluble in supercritical carbon dioxide or sulfur hexafluoride. Derivatization/supercritical fluid extraction (D/SFE) is the derivatization of a chemical moiety to a less polar form which can be extracted by SFE. Lyophilized *Escherichia coli* cell mass was treated with the reagent trimethylphenyl ammonium hydroxide to derivatize polar lipid fatty acids to their methyl esters prior to extraction by SFE (*58*). Figure 1 is a comparison of the PLFA profiles generated by D/SFE with the solvent extraction, silicic acid fractionation, and derivatization method (*59*). The reduction in the amount of time required for sample preparation by this method is considerable - by supercritical fluid chromatography 8 samples can be prepared and analyzed by GC in one day, while by solvent extraction the same samples would require about 3 days to prepare.

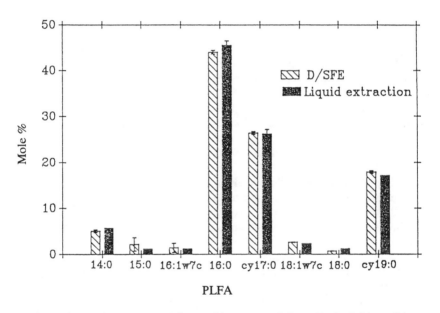

Figure 1. Mole percents of fatty acids recovered from *Escherichia coli* by liquid extraction, fractionation, and derivatization compared with those recovered by derivatization/supercritical fluid extraction (D/SFE).

Applications of Rapid Lipid Biomarker Analysis

Rapid Diagnosis of Acute Infections. The rapid and accurate diagnosis of infectious disease is critical to appropriate patient management (*60*) as well as epidemiologic surveillance (*61*). Isolation and culturing of infectious agents was essential in the definition of the etiology of specific infectious diseases and still represent the definitive diagnostic criteria. The problem with this approach is that the procedures require time to isolate the organism and then to grow it on specific media. The media must be chosen judiciously based on the clinical situation since many pathogens require specific growth conditions and nutrients (*62*). There is also an increasing recognition that there are infections with agents that are "viable but not culturable" (*63*).

There has been great progress in rapid diagnosis of specific microorganisms that is based on DNA probes (*64, 65*), rRNA probes (*5*), and polyclonal or monoclonal antibodies (*66*). In each case, however, the nature of the pathogen(s) must be suspected to choose the proper probes. Some of the probes are too specific and miss different strains of the pathogen and in some cases the probes are too general.

Examples of potential marker lipids and their associated disease organisms include: tuberculostearic acid from *Mycobacterium tuberculosis* (*vide supra, 39*), ergosterol from the yeast *Candida* spp. (*33*), and the distinctive PLFA and quinones of *Legionella pnuemophila* (*67*).

Monitoring and Predicting Bioremediation Strategies. Signature lipid biomarker analysis can be utilized for monitoring and control of bioremediation systems. For example, aerobic trichloroethylene biodegradation is a co-metabolic process of either methanotrophs consuming methane, or alkanotrophs consuming propane. Subsurface sediment bioremediation has been found to have the highest rate when the bacteria are nutritionally stressed, as indicated by a high ratio of poly-ß-hydroxyalkanoates to PLFA (*68*). PLFA profiles have also been used to monitor the populations of methane (*69*) or propane (*70*) oxidizing bacteria in bioreactors and subsurface sediments.

Environmental Toxicity Assessment. Prediction of the effects of industrial or municipal effluents on natural systems is complicated by the mixture of components in any real waste stream and the multitude of organisms composing most ecosystems. Methods proposed or applied to monitoring the effects of putative toxins on natural environments include quantification of the toxic compounds (*71*), and physiologic functions such as reproductive success (*72*). Application of PLFA analysis to stream periphyton communities (*73*) at 3 sites with sequentially less organic and metals contamination showed that the microbial community structure had been strongly impacted by effluents entering the stream, in agreement with fish liver function biomarkers, and algal periphyton density and physiological stress. However, little evidence was found for acute or chronic toxicity using either the *Ceriodaphnia* or fathead minnow bioassays.

Acknowledgments

We would like to recognize Tina Anderson for her work on the supercritical fluid extraction and the derivatization/supercritical fluid extraction systems.

Literature Cited

1. Welch, D. F. *Clin. Microbiol. Rev.* **1991**, *4*, 422-438.
2. Stahl, D. A.; Flesher, B.; Mansfeld, H. R.; Montgomery, L. *Appl. Environ. Microbiol.* **1988**, *54*, 1079-1084.
3. Steffan, R. J.; Atlas, R. M. *Ann. Rev. Microbiol.* **1991**, *45*, 137-161.
4. Jiménez, L. *Appl. Environ. Microbiol.* **1990**, *56*, 2108-2113.
5. Stahl, D. A.; Flesher, B.; Mansfield, H. R.; Montgomery, L. *Appl. Environ. Microbiol.* **1988**, *54*, 1079-1084.
6. Kobayashi, H. A.; De Macario, E. C.; Williams, R. S.; Macario, A. J. L. *Appl. Environ. Microbiol.* **1988**, *54*, 693-698.
7. Nivens, D. E.; Chambers, J. Q.; White. D. C. In *Microbially Influenced Corrosion and Biodeterioration;* Dowling, N. J. E.; Mittelman, M. W.; Danko, J. C., Ed.; University of Tennessee: Knoxville, Tennessee, 1991; pp 5-47.
8. Peck, M. W.; Chynoweth, D. P. *Biotechnol. Letts.* **1990**, *12*, 17-22.
9. Brasswell, S. C.; Eglinton, G. In *Organic Marine Geochemistry, ACS Symp. Ser. 305*; Sohn, M. L., Ed.; American Chemical Society: Washington, D. C., 1986; pp 11-32.
10. Bligh, E. G.; Dyer, W.J. *Can. J. Biochem. Biophysiol.* **1959**, *35*, 911-917.
11. Clement, R. E.; Karasek, F. W. In *Mass Spectroscopy in the Environmental Sciences*; Karasek; F. W.; Hutzinger, O.; Safe, S., Eds.; Plenum Press: New York, NY, 1985; pp. 21-48.
12. Smith, R. D.; Wright, B. W.; Kalinoski, H. T. *Progress in HPLC* **1989**, *4*, 111-155.
13. Pinkston, J. D.; Delaney, T. E.; Bowling, D. J.; Chester, T. L. *J. High Resol. Chromatog.* **1991**, *14*, 401-406.
14. Alexander, M. *Introduction to Soil Microbiology*, 2nd ed., John Wiley & Sons, New York, 1977.
15. Roszak, D. B.; Colwell, R. R. *Microbiol. Rev.* **1987**, *51*, 365-379.
16. Ferguson, R. L.; Rublee, P. *Limnol. Oceanogr.* **1976**, *21*, 141-145.
17. Balkwill, D. L.; Ghiorse, W. C. *Appl. Environ. Microbiol.* **1985**, *50*, 580-588.
18. Olsen, R. A.; Bakken, L. R. *Microb. Ecol.* **1987**, *13*, 59-74.
19. Kates, M. *Adv. Lipid Res.* **1964**, *2*, 17-90.
20. Smith, G. A.; Nickels, J. S.; Kerger, B. D.; Davis, J. D.; Collins, S. P.; Wilson, J. T.; McNabb, J. F.; White, D. C. *Can. J. Microbiol.* **1986**, *32*, 104-111.
21. White, D. C.; Davis, W. M.; Nickels, J. S.; King, J. D.; Bobbie, R. J. *Oecologia*, **1979**, *40*, 51-62.
22. Woese, C. R.; Kandler, O.; Wheelis, M. *Proc. Natl. Acad. Sci. USA* **1990**, *87*, 4576-4579.
23. Balkwill, D. L.; Leach, F. R.; Wilson, J. T.; McNabb, J. F.; White, D. C. *Microb. Ecol.* **1988**, *16*, 73-84.

24.   Tornebene, T. G.; Langworthy, T. A. *Science* **1979**, *293*, 51-53.
25.   DeLuca, S.J.; Voorhees, K. J.; Langworthy, T. A.; Holzer, G. *Biochim. Biophys. Acta* **1986**, *9*, 487-492.
26.   Hedrick, D. B.; Guckert, J. B.; White, D. C. *J. Lipid Res.* **1991**, *32*, 659-666.
27.   Hedrick, D. B.; Richards, B.; Jewell, W.; Guckert, J. B.; White, D. C. *J. Indust. Microbiol.* **1991**, *8*, 91-98.
28.   Hedrick, D. B.; White, T.; Guckert, J. B.; Jewell, W. J.; White, D. C. *J. Indust. Microbiol.* **1992**, *9*, 193-199.
29.   Hedrick, D. B.; Pledger, R. J.; White, D. C.; Baross, J. A. *FEMS Microbial Ecology* **1992**, *101*, 1-10.
30.   Mancuso, C. A.; Franzmann, P. D.; Burton, H. R.; Nickels, P. D. *Microbial Ecol.* **1990**, *19*, 73-95.
31.   Dickins, H. D.; Vanvleet, E. S. *Deep-Sea Research, Part A*, **1993**, *39*, 521-536.
32.   Pauly, G. G.; Van Vleet, E. S. *Geochim. Cosmochim. Acta* **1986**, *50*, 1117-1125.
33.   Weete, J. D. *Lipid biochemistry of fungi and other organisms;* Plenum Press: New York, NY, 1980.
34.   White, D. C.; Bobbie, R. J.; Nickels, J. S.; Fazio, S. D.; Davis, W. M. *Botanica Marina* **1980**, *23*, 239-250.
35.   Nichols, P. D.; Shaw, P. M.; Johns, R. B. *J. Microbiol. Meth.* **1985**, *3*, 311-320.
36.   Nichols, P. D.; Mayberry, W. R.; Antworth, C. P.; White. D. C. *J. Clin. Microbiol.* **1985**, *21*, 738-740.
37.   Nichols, P. D.; Smith, G. A.; Antworth, C. P.; Hanson, R. S.; White. D. C. *FEMS Microbiol. Ecol.* **1985**, *31*, 327-335.
38.   Kerger, B. D.; Nichols, P. D.; Antworth, C. P.; Sand, W.; Bock, E.; Cox, J. C.; Langworthy, T. A.; White. D. C. *FEMS Microbiol. Ecol.* **1986**, *38*, 67-77.
39.   Larsson, L.; Odham, G.; Westerdahl, G.; Olsson, B. *J. Clin. Microbiol.* **1987**, *25*, 893-896.
40.   Dobbs, F. C.; Guckert, J. B. *Marine Ecol. Prog. Ser.* **1988**, *45*, 69-79.
41.   Ratledge C.; Wilkinson. S. G. *Microbial Lipids* Plenum Press: New York, NY, 1988; Vol 1.
42.   Collins, M. D.; Jones, D. *Microbiol. Rev.* **1981**, *45*, 316-354.
43.   Hedrick, D. B.; White, D. C. *J. Microbial Meth.* **1986**, *5*, 49-55.
44.   Kroppenstedt, R. M. *J. Liquid Chrom.* **1982**, *5*, 2359-2367.
45.   Gere, D. R. *Science* **1983**, *222*, 253-258.
46.   Lepage, C.; Fayolle, F.; Hermann, M.; Vandecasteele, J.-P. *J. Gen. Microbiol.* **1987**, *133*, 103-110.
47.   Guckert, J. B.; Hood, M. A.; White, D. C. *Appl. Environ. Microbiol.* **1986**, *44*, 794-801.
48.   Nickels, J. S.; King, J. D.; White, D. C. *Appl. Environ. Microbiol.* **1979**, *37*, 459-465.
49.   Findlay, R. H.; White, D. C. *Appl. Env. Microbiol.* **1983**, *45*, 71-78.
50.   Guckert, J. B.; Cooksey, K. E. *J. Phycol.* **1990**, *26*, 72-79.
51.   Gehron, M. J.; White, D.C. *J. Exp. Mar. Biol.* **1982**, *64*, 145-158.
52.   Chester, T. L. *J. Chrom.* **1984**, *299*, 424-431.

53. Lipsky, S. R.; Duffy, M. L. *J. High Resolution Chrom. & Chrom. Comm.* **1986,** *9,* 725-730.
54. *Microbial Identification System Operating Manual, Ver. 3.* Microbial ID, Inc.: Newark, Delaware.
55. Regnery, R. L.; Anderson, B. E.; Clarridge, J.E.; Rodriguez-Barradas, M. C.; Jones, D. C.; Carr, J. H. *J. Clin. Microbiol.* **1992,** *30,* 265-273.
56. Gitaitis, R. D.; Sasser, M.; Beaver, R. W.; McInnes, T. B.; Stall, R. E. *Phytopathology* **1987,** *77,* 611-615.
57. Mirza, M. S.; Janse, J. D.; Hahn, D.; Akkermans, A. D. L. *FEMS Microbiol. Ecol.* **1991,** *83,* 91-98.
58. Hawthorne, S. B.; Miller, D. J.; Nivens, D. E.; White, D. C. *Anal. Chem.* **1992,** *64,* 405-412.
59. Nivens, D. E.; Anderson, T.; White, D. C. In preparation.
60. Evans, A. S. In *Bacterial Infections of Humans: Epidemiology and Control*; Evans, A. S.; Brachman, P. S., Eds.; Plenum Press: New York, NY, 1991; pp. 3-58.
61. Brachman, P. S. In *Bacterial Infections of Humans: Epidemiology and Control*; Evans, A. S.; Brachman, P. S., Eds.; Plenum Press: New York, NY, 1991; pp. 59-74.
62. Sommers, H. M. In *Tuberculosis*; Youmans, G. P., Ed.; Saunders Co.: Philadelphia, PA, 1979; pp. 404-434.
63. Colwell, R. R.; Brayton, P. R.; Grimes, D. J.; Rozak, D. B.; Huq, S. A.; Palmer. L. M. *Biotechnology* **1985,** *3,* 817-820.
64. Tenover, F. C. *Clin. Microbiol. Rev.* **1988,** *1,* 82-101.
65. Pierre, C.; Olivier, C.; Lecossier, D.; Boussougant, Y.; Yeni, P. *Am. Rev. Respiratory Dis.* **1993,** *147,* 420-424.
66. Ørskov, I.; Ørskov, F.; Jann, B.; Jann, K. *Bacteriol. Rev.* **1977,** *41,* 667-710.
67. Ehret, W.; Jacob, K.; Ruckdeschel, G. *Zbl. Bakt. Hyg. A* **1987,** *266,* 261-275.
68. Nichols, P. D.; White, D. C. *Hydrobiologia* **1989,** *176/177,* 369-377.
69. Nichols, P. D.; Smith, G. A.; Antworth, C. P.; Parsons, J.; Wilson, J. T.; White, D. C. *Environ. Tox. Chem.* **1987,** *6,* 89-97.
70. Ringelberg, D. B.; Davis, J. D.; Smith, G. A.; Pfiffner, S. M.; Nichols, P. D.; Nichels, J. B.; Henson, J. M.; Stocksdale, T. T.; White, D. C. *FEMS Microbiol. Ecol.* **1988,** *62,* 39-50.
71. Davies, I. M.; McKie, J. C.; Paul, J. D. *Aquaculture* **1986,** *55,* 103-114.
72. Gillespie, R. B.; Baumann, P. C. *Trans. Am. Fish. Soc.* **1986,** *115,* 208-213.
73. Guckert, J.B., Nold, S.C., Boston, H.L., and White, D.C. *Can. J. Fish. Aquatic Sci.* **1992,** *49,* 2579-2587.

RECEIVED April 16, 1993

Chapter 3

Fast Atom Bombardment Mass Spectrometry of Phospholipids for Bacterial Chemotaxonomy

D. B. Drucker

Department of Cell and Structural Biology, School of Biological Sciences, University of Manchester, Manchester M13 9PT, England

Fast atom bombardment mass spectrometry (FAB-MS) permits analysis of bacterial polar lipids and provides data of taxonomic value. Members of the phospholipid series phosphatidylserine, phosphatidylglycerol and phosphatidylethanolamine are especially well seen as their anions. Members of the enterobacteria have qualitatively similar anionic polar lipid fingerprints which differ considerably from less closely related bacteria, e.g. *Acinetobacter*, *Pseudomonas*, *Bacillus*, *Staphylococcus*, *Micrococcus*, *Streptococcus*, *Lactobacillus*, *Capnocytophaga* and *Prevotella*. These genera have polar lipid profiles that also differ from one another, qualitatively. FAB-MS is thus able to differentiate qualitatively between many species. Use of numerical analysis of normalised quantitative data also permits objective comparison of closely related strains which have similar polar lipid profiles for purposes of classification and identification.

Chemotaxonomy, or chemosystematics, is the use of chemical analytical data for classification of microorganisms. Once such a classification scheme has been created, it may provide information of use in identifying single, unidentified, strains. Sometimes chemical data alone can be used, on other occasions they are used along with other information such as Gram-staining reaction, cellular morphology, and biochemical reactions.The types of chemical data used are described elsewhere (*1*) but have frequently employed gas chromatographic (GC) analysis (*2*). Because GC is a separative technique primarily, its combination with mass spectrometry has provided surer peak identification for sugars (*3*), carboxylic methyl esters (*4*), and amines (*5*). The use of GC introduces variability due to column and operating temperatures which can be circumvented by use of (Figure 1) fast-atom bombardment mass spectrometry (FAB-MS) which effectively separates mixtures of compounds such as polar lipids (*6*) without the need for prior GC separation. Polar lipids are particularly well seen in anionic spectra and their extraction from cells, with hydrophobic solvents, effectively removes them from hydrophilic cell constituents. Their surface activity also ensures that they are well suited to ionization by xenon from their location in a matrix fluid, such as glycerol or 2-nitrobenzyl alcohol.

0097–6156/94/0541–0018$06.00/0
© 1994 American Chemical Society

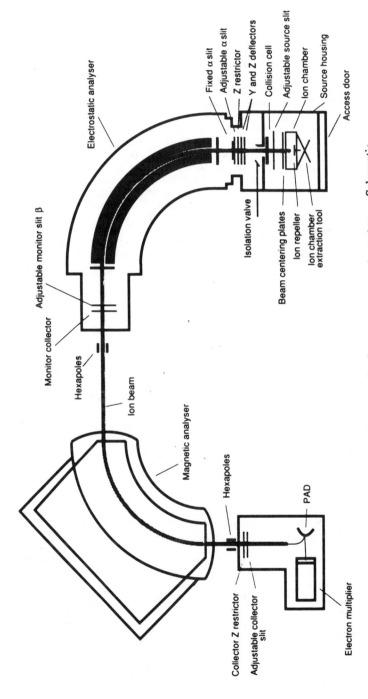

Figure 1. Fast atom bombardment mass spectrometry. Schematic diagram of Kratos Concept IS (reproduced with permission of Kratos Analytical).

Bacterial Polar Lipids

With regard to FAB-MS, polar lipids are taken to include such compounds as phospholipids and their fatty acid constituents. However, fatty acids arise in spectra from polar lipid breakdown and aid interpretation of lipid anion peaks (*vide infra*). In this chapter, phospholipids will primarily be considered. Typical bacterial phospholipids are shown in Figure 2. The phospholipids characteristically contain a glycerol-phosphate core with two fatty acyl substituents and either glycerol, ethanolamine or serine, to give respectively the phospholipids phosphatidylglycerol, phosphatidylethanolamine or phosphatidylserine, all detected as anions. When phosphatidylcholine is present, the protonated ion is well seen in positive-ion spectra. The fatty acyl substituents usually range from C_{12} to C_{23}. They may be mono- or di- unsaturated, contain cyclopropane rings, or bear hydroxy substituents. Thus 'phosphatidylglycerol' is a family of hundreds of compounds with slightly different R^1 and R^2 (fatty acyl) substituents. The historical approach to analysis of bacterial polar lipids has been to use TLC separation, which merely separates the major classes, or to use HPLC which with skill can provide detailed structural information by comparison of data with retentions of standards (*7*). In addition to the lipids shown in Figure 2, many others are known including 'ornithine-containing lipids (*8*), lipid A (*9*), archebacterial ether lipids (*10*), aminophospholipids (*11*), rhamnolipids (*12*), as well as lysylphosphatidylglycerol, diphosphatidylglycerol, diglycosyl-diglyceride and phophatidylinositol (*13*).

FAB-MS of Bacteria

The initial discovery that argon could be used to bombard surfaces and solids to create ions for mass spectral analysis (*14*) led to the biochemical use of FAB-MS (*15*). A variety of studies subsequently employed the technique for examination of bacteria. FAB-MS was used to examine acidic polysaccharides of *Rhizobium* (*16*), endotoxins(*17*), aminolipids(*11*), poly 3-hydroxybutanoate (*18*), and a disaccharide determinant of *Staph. aureus(19)*. The technique has also been applied to the study of protein primary sequences (*20*), protein N-terminal analysis (*21*), bacterial toxins, *e.g. E.coli* heat-stable enterotoxin II (*22*), and endotoxin(*23,17*). FAB-MS has proved of particular value in examining polar lipids of bacteria for which it provides a simple means of analyzing specific moieties not readily analyzed so rapidly by alternative means such as phospholipid, which have required skilled and intensive study by HPLC (*7*).

FAB-MS of Polar Lipid

Phosphatidylcholine. Several studies have now examined pure polar lipids by FAB-MS. An early study (*24*) analyzed lecithins from egg yolk and soybean which were compared with specific phosphatidylcholines, phosphatidylethanolamines, dipalmitoylphosphatidylserine and dipalmitoylphosphatidyl glycerol. Samples were dissolved in glycerol matrix, bombarded with argon, and positive ion (PI) FAB spectra recorded. It was observed that...'abundant molecular ions are noted together with fragment ions which provide structural information about the acyl and phosphate ester groupings'. One bonus afforded by FAB-MS was that only simple sample preparation procedures were required.

In PI spectra, phosphatidylcholines yielded stronger peaks than the other phospholipids tested. In the case of dilauryl phosphatidylcholine (mol. wt. 621), a protonated ion was observed at m/z 622. Other significant ions were at m/z 440 and m/z 424; the former represents loss of an acyl chain $[MH-COC_{11}H_{22}]^+$, the latter arises through loss of carboxylate $[MH-OCOC_{11}H_{22}]^+$. Other ions generally observed were at m/z 184 [choline phosphate] and m/z 86[CH_2=CH-$N^+(CH_3)_3$]. Characteristically, the predicted ions were accompanied by less intense ions two daltons below the expected m/z values.

Surprisingly, this study (24) claimed that neither the ubiquitous phosphatidylcholine nor many other phospholipids (e.g. phosphatidylethanolamine) 'showed any molecular, or fragment ions under negative ion (NI) FAB conditions'. A subsequent major study (25) found that it was possible to analyze phosphatidylcholine by NI-FAB-MS. Most importantly, such spectra could be subdivided into four regions, viz. molecular ion , intermediate mass, carboxylic acid residue, and a lower mass region. An understanding of the identity and source of the ions of lower m/z values is invaluable for correct identification of the higher m/z values due to polar lipids. Ions observed with dipalmitoyl phosphatidylcholine (mol. wt. 733) were at m/z 718(loss of $[CH_3]^+$), m/z 673(loss of $[CH_2N(CH_3)_2]$), and the most intense peak was at m/z 255 (hexadecanoate). The elimination of CH_3 from the molecular ion forms a quasimolecular ion indistinguishable from the corresponding dimethylethanolamine. This led Muenster et al. (25) to conclude that...'in this case, PI-FAB has to be resorted to where $[M+H]^+$ retains the [-$N(CH_3)_3]^+$ group of choline'. A recent study (26) has used both PI and NI-FAB-MS for structural analysis of choline phospholipids. Dipalmitoylglycerylphosphatidyl choline yielded cations of m/z 734 which are due to $[M+H]^+$. Loss of choline phosphate provided a peak at m/z 184 (*vide supra*) which left the ion $[M-184]^+$ of m/z 550. Other peaks in the intermediate or deacylated region were of m/z 496 $[M-238]^+$ and of m/z 494. In negative-ion mode, the peak of m/z 718 was again observed (25) but two other peaks, viz. m/z 673 and 647, were noted as components of a 'triplet-ion', characteristic of phosphatidylcholine in NI-FAB-MS. The peak of higher m/z is equivalent to $[M-CH_3]^-$ as has been explained; similarly, the peak of lower m/z is equivalent to $[M-HN(CH_3)_3]^-$ and the lowest can be assigned to $[M-CH_2=CHN(CH_3)_3]^-$. The triplet ion is also characteristic of lysophosphatidylcholine(26). (See Table I.)

Table I. Interpretation of NI-FAB-MS of Phosphatidylcholine

Anions	m/z values
Major peaks	
[Phospholipid-CH_3]$^-$	M-15
[R^1COO]$^-$ carboxylate	M^1-1
[R^2COO]$^-$ carboxylate	M^2-1
Minor peaks	
[M-$CHN(CH_3)_3$]$^-$	M-60
[M-CH_2=$CHN(CH_3)_3$]$^-$	M-86

M, M^1, M^2, are mol. wt. of PC, and carboxylic acids R^1COOH and R^2COOH.

Phosphatidylethanolamine

PI-FAB-MS. Several studies have examined pure samples of phosphatidylethanolamine(PE) by FAB-MS (27,24,25,28). Initially, PI ion spectra were used for FAB-MS analysis of PE (24). In the case of dipalmitylglycerylphosphatidyl ethanolamine, the $[MH]^+$ ion observed is of m/z 692, 706, or 720 according to whether the head group is unmethylated, N-methylated, or N,N-dimethylated. Diagnostically important ions are for $[MH-2]^+$, and derived by deacylation by alpha-cleavage and hydrogen rearrangement, e.g. of m/z 454 $[MH-COC_{15}H_{30}]^+$; loss of carboxylate accounts for the ion of m/z 438 $[MH-OCOC_{15}H_{30}]^+$. PE also yields diagnostic base peaks of m/z 170, 156, or

142 from DMPE, MMPE, or PE respectively which are due to ethanolamine phosphate ions. One problem associated with PI-spectra is that peaks are observed for $[M+Na]^+$ and $[M+K]^+$ which can complicate interpretation of mass spectra (28).

NI-FAB-MS. NI-spectra of PE (25) yield data analogous to those for PC (*vide supra*). A characteristic feature of spectra is the triplet ion corresponding to a major peak due to $[M-H]^-$ which is accompanied by smaller peaks of m/z 17 and 43 integers less than the molecular anion due to $[M-H-NH_3]^-$ and {M-$CH_2CH_2NH_2$}. Additional fragmentation provides a carboxylate ion $[RCOO]^-$ which is usually the base peak and leaves a fragment equivalent to loss of carboxylate $[M-H-RCOOH]^-$. In the case of phosphatidylethanolamine, little further fragmentation occurs unlike PC. However, a small peak can arise from loss of choline phosphate $[M-PO_4CH_2CH_2N(CH_3)_3]^-$ with a peak of m/z 183 less than that of the molecular anion. In the lower mass region, peaks are observed of m/z 180 and 140 which are absent from spectra for phosphatidic acid with identical fatty acyl groups. Such peaks are considered diagnostic (25) for an ethanolamine head group. Nevertheless, the most intense peaks are due to $[M-H]^-$ and $[RCOO]^-$. (See Table II.)

Table II. Interpretation of NI-FAB-MS of PE

Anions	m/z values
Major peaks	
[Phospholipid-H]$^-$	M-1
[R^{1}COO]$^-$ carboxylate	M^1-1
[R^{2}COO]$^-$ carboxylate	M^2-1
Minor peaks	
[M-H-NH$_3$]$^-$	M-18
[M-CH$_2$CH$_2$NH$_2$]$^-$	M-44
[M-H-R^{1}COOH]$^-$	M-1-M^1
[M-H-R^{2}COOH]$^-$	M-1-M^2
[M-PO $_4$CH$_2$CH$_2$N(CH$_3$)$_3$]$^-$	M-183

M, M^1, M^2 correspond to mol. wt. of PE, and two fatty acids, R^{1}COOH and R^{2}COOH.

Phosphatidylglycerol

PI-spectra. An early study (24) found that PI-FAB-MS yielded diagnostically important ions from PG. These corresponded to $[M+H]^+$, [MH-2]$^+$, $[M-237]^+$, $[M-239]^+$, $[M-171]^+$ and $[M-409]^+$ when dipalmitoylphosphatidylglycerol was examined. The $[M-239]^+$ peak might represent loss of -COC$_{15}$H$_{31}$ from the $[M+H]^+$ cation.

NI-spectra. Initial study (24) suggested that only PG yielded PI-FAB-MS peaks, with a major peak for dipalmitoylPG of m/z 721 which is equivalent to $[M-H]^-$. Later work (25) found that the same compound yields peaks of m/z 721 (most intense peak) and of 647, 483, 465, 409, 391, 255, 211 and 171. The lack of a peak of equivalent to $[M-18]^-$ was considered to be a diagnostic feature (*vide supra*). One obvious difference from PE is that PG has an anion of odd m/z value whereas PE which contains a nitrogen atom has an anion of even m/z value. (See Table III.)

Table III. Interpretation of NI-spectra of PG

Anions	*m/z* values
Major peaks	
Phospholipid	M-1
Carboxylate	M^1-1
	M^2-1
Minor peaks	
Unassigned ion	M-75
Unassigned ion	M-239
[M-R^{1}COOH-H]⁻	M-1-M^1
[M-R^{2}COOH-H]⁻	M-1-M^2
Unassigned ion	M-313
Unassigned ion	M-511
Unassigned ion	M-551

M, M^1, M^2, represent mol. wt. of PG, and two carboxylic acids, R^1COOH and R^2COOH respectively.

Phosphatidylserine. PS has been studied, along with other phospholipids, by PI-FAB-MS. One study (*3*) reported that FAB…'provided protonated ions in all cases with the exception of phosphatidylserines'. Another study (*24*) found that an ion of m/z 496 was present but no other fragmentations. An ion of m/z could be attributed to loss of phosphate ester. Little is known of NI-FAB-MS of PS.

FAB-MS of Microbial Polar Lipid

Algae. *Scenedesmus obliquus* and *Chlorella vulgaris* have been examined both by NI- and PI-FAB-MS (*29*). Samples consisted of either '1-10 mg intact dried algal cells' or a '1-3 mg ml⁻¹ crude lipid extract' solution dissolved in triethanolamine. In NI-FAB-MS, ions derived from the matrix were observed at m/z 297 and m/z 148, corresponding to [2 TEA-H]⁻ and [TEA-H]⁻ respectively. Major peaks were observed in the carboxylate anion region of m/z 297 which corresponds to nonadecanoate, and of m/z 281, 279, 255, and 253, identified as octadecenoate, octadecdienoate, hexadecanoate, and hexadecenoate anions. Cells and lipid extracts yielded similar results. Data for parent polar lipids were not provided.

Enterobacteria. These are Gram-negative rod-shaped bacteria capable of growing in the presence of bile and facultatively with respect to oxygen, *i.e.* in its presence or absence.

    Escherichia coli has been examined in a number of FAB-MS studies (*28,13,30*). Only one study has employed PI-FAB-MS (*28*) and has reported major anions of m/z 563, 466, 591, 704, 705, and 732 among others (Figure 3). Peaks of m/z 704 and 732 were regarded as protonated PE/33:1 and PE/35:1 where the acyl substituents were hexadecanoate with heptadecenoate and nonadecenoate respectively. In reality, the monoenoates may well have been cyclopropane acids (*30*). Peaks in the intermediate region would have been due to corresponding diacylglycerols. One confusing aspect of PI spectra is the appearance of peaks attributed to salt complexes. A peak of m/z 726 was attributed

a) CH_2-OCOR^1
    |
    $CH-OCOR^2$
    | O
    | ||
    $CH_2-OPO-CH_2CH_2N^+(CH_3)_3$
         |
         O^-

d) CH_2-OCOR^1
    |
    $CH-OCOR^2$
    | O
    | ||
    $CH_2-OPO-CH_2CH$ CH_2OH
         | |
         OH   OH

b) CH_2-OCOR^1
    |
    $CH-OCOR^2$
    | O
    | ||
    $CH_2-OPO-CH_2CH_2NH_2$
         |
         OH

e) CH_2-OCOR^1
    |
    $CH-OCOR^2$
    | O
    | ||
    CH_2-OPO
         |
         OH

c) CH_2-OCOR^1
    |
    $CH-OCOR^2$
    | O
    | ||
    $CH_2-OPO-CH_2CHCOOH$
         | |
         OH NH_2

Figure 2.   Phospholipid   structures.   a)phosphatidylcholine,
b)phophatidylethanolamine, c)phosphatidylserine, d)phosphatidylglycerol,
e)phosphatidylinositol

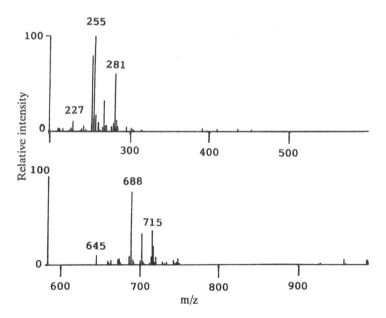

Figure 3.   PI-FAB mass spectrum of polar lipids of *Escherichia coli*,
adapted from (*28*).

Table IV. Major Anions in NI-FAB-MS Spectra of Phosopholipid of Various Bacteria

Bacterial species	Major anions
Enterobacteriaceae	
C.freundii	688 > 702 > 716 > 689 > 714 > 717 > 719 > 747 > 715
Ent.cloacae(012)	688 > 716 > 714 > 703 > 689 > 686 > 702 > 645 > 662
Ent.cloacae(008)	688 > 703 > 716 > 714 > 689 > 690 > 717 > 686 > 720
E.coli(015)	688 > 716 > 702 > 689 > 714 > 717 > 719 > 747 > 703
E.coli(017)	702 > 716 > 688 > 747 > 703 > 714 > 719 > 733 > 717
E.coli(014)	702 > 688 > 716 > 747 > 719 > 714 > 717 > 689 > 733
E.coli*	702 > 730 > 703 > 716 > 731 > 733 > 747
M.morganii(020)	702 > 733 > 703 > 688 > 734 > 674 > 716 > 659 > 719
Pl.shigelloides(013)	688 > 689 > 716 > 702 > 714 > 686 > 719 > 717 > 690
Pr.mirabilis(010)	688 > 703 > 720 > 716 > 717 > 748 > 690 > 691 > 689
Pr.mirabilis(001)	702 > 688 > 716 > 730 > 703 > 689 > 717 > 704 > 663
Pr.mirabilis(016)	688 > 716 > 702 > 719 > 689 > 747 > 717 > 690 > 703
Pr.vulgaris*	702 > 703 > 674 > 688 > 716
Ser.liquefaciens(019)	688 > 702 > 689 > 716 > 703 > 719 > 690 > 662 > 714
Ser.marcescens(003)	702 > 688 > 689 > 703 > 704 > 662 > 686 > 690 > 700
NF Gram-ve rods	
Acinetobacter sp. (011)	717 > 748 > 744 > 747 > 743 > 714 > 716 > 774 > 689
Acinetobacter sp. (004)	748 > 716 > 743 > 718 > 714 > 717 > 745 > 747 > 674
Ps.fluorescens*	702 > 703 > 688 > 716 > 733
Anaerobic Gram-negative rods	
Prev.intermedia	663 > 692 > 706 > 664 > 649 > 678 > 635 > 661 > 677
Prev.melaninogenica	589 > 590 > 587 > 603 > 663 > 654 > 588 > 619
Capnophilic Gram-negative rods	
C.gingivalis	574 > 588 > 662 > 575 > 618 > 572 > 619 > 653 > 589
Gram-positive cocci	
M.luteus*	693 > 694 > 707 > 766 > 768 > 781 > 679 > 721
Staph.aureus*	721 > 735 > 763 > 749 > 723 > 736 > 764 > 693
Strep.mutans	747 > 688 > 716 > 1077 > 749
Non-sporing Gram-positive Rods	
L.rhamnosus	761 > 762 > 759 > 787 > 733 > 760 > 763 > 788 > 747
Sporing Gram positive rods	
B.subtilis	721 > 693 > 722 > 694 > 707 > 690 > 691 > 708

Data from (13,32,* and 30) or unpublished data.
Major peaks shown.

to [PE/33:1 + Na]$^+$. NI-spectra (Table IV) are perhaps more straightforward to interpret. Major peaks are of m/z 688, 702, 716, 689, 747, 714, and 703 in one study (31) and peaks of m/z 702, 730, 703, 716, and 731 in another study (13). In the former study, peaks corresponded to PE/32:1, PE/33:1, PE/34:1, PG/30:2, PG/34:1, PG/34:2, and PG/31:2. In the latter study, additional peaks of 730 and 716 can be identified as PE/35:1, and PE/34:1 which have been reported in another study (30). These differences almost certainly reflect strain differences that exist within *E.coli*. Other enterobacteria have also been studied by FAB-MS, although less extensively.

Citrobacter freundii. This has been examined in one study (*31*) and by NI-FAB-MS (Table IV). The major anions are virtually identical to those of a non-lactose fermenting *E.coli* strain. Indeed, *Citrobacter* is very closely related to *Escherichia*. The major phospholipids then are of m/z 688 (PE/32:1), 702 (PE/33:1), 7I6(PE/34:1), 689 (PG/30:2) and 714 (PE/34:2).

Enterobacter cloacae. This has, again, been examined in only two studies (*31*) and NI spectra (Figure 4) reveal peaks similar to those of other enterobacteria. Major phospholipid peaks (Table IV) are of m/z 688 (PE/32:1), 716(PE/34:1), 703 (PG/31:2), 714 (PE/34:2) and 689 (PG/30:2).

Morganella morganii. This has been examined in one study (*31*) and the major peaks (Table IV) are of m/z 702(PE/33:1), 733 (PG/33:1), 703 (PG/31:2), 688 (PE/32:1) and 734 (PE/OH-34:0). Of some interest is the peak of m/z 734.

Plesiomonas shigelloides. The same study (*31*) has examined this enterobacterial strain (Table IV) which has peaks of m/z 688 (PE/32:1), 689(PG/30:2), 716(PE/34:1), 702 (PE/33:1) and 714 (PE/34:2).

Proteus. *Proteus mirabilis* has been examined in three studies (*28,30,31*) and *Proteus vulgaris* in one (*13*). Data are shown in Table IV. For *Pr.mirabilis* major peaks (Figure 5) are found of m/z 688 (PE/32:1), 702(PE/33:1). 716 (PE/34:1), 730 (PE/35:1), 703 (PG/31:2), 719 (PG/32:1) and 689 (PG/30:2). For *Pr.vulgaris*, major peaks are of m/z 702, 703, 674 (PE/31:1), 688 and 716. The peak of m/z 674 is also present, though not as a major peak, in *Pr.mirabilis* (*31*).

Serratia. Two species have been examined (*31*) and major peaks are shown in Table IV. Major peaks in both *S.marcescens* and *S.liquefaciens* are of m/z 702 (PE/33:1), 688 (PE/32:1), 689 (PG/30:2), and 703 (PG/31:2). In *S.marcescens*, a peak of m/z 704 (PE/33:0) is also a major peak. Similarly a peak of m/z 716 (PE/34:1) is a major peak in *S.liquefaciens*.

NF Gram-ve Rods. This group of microorganisms are non-fermenting Gram-negative rods incapable of growing under anaerobic (oxygen-free) conditions.Species examined include *Pseudomonas aeruginosa* (*28*), *Ps.fluorescens* (*13*) and *Acinetobacter calcoaceticus* (*31*).

    Pseudomonas. All have been examined by NI-FAB-MS although the spectrum for *Ps.aeruginosa* (*28*) appears to be a PI spectrum with m/z values corresponding to [M+H]$^+$ for PE and PC. Major phospholipid anion peaks are of m/z 718 (palmityloleylphosphatidyl ethanolamine), m/z 760 (palmityloleylphosphatidyl choline), m/z 716 (palmitoleyloleylphosphatidyl ethanolamine), m/z 824 and m/z 692 (dipalmitylphosphatidyl ethanolamine). Peaks at m/z 577 and 575 can be ascribed to loss of ethanolamine phosphate which gives rise to an anion of m/z 142 and leaves a diacyl glyceride with $C_{16:0}/C_{18:1}$ and $C_{16:1}/C_{18:1}$ substituents. NI-FAB-MS spectra have been published elsewhere for *Ps.fluorescens* (*13*) and major peaks are listed in Table IV , viz. m/z 702, 703, 688, 716, 733. These values correspond to PE/33:1, PG/31:2, PE/32:1. PE/34:1 and PG/33:1 respectively.

    Acinetobacter calcoaceticus. This has also been examined by NI-FAB-MS (*31*) and major peaks are listed in Table IV. The most abundant ions are of m/z 717, 748, 744, 747, 743, 718 and 714. These correspond to PG/32:2, PS/33:0, PS/33:2, PG/34:1, PG/34:3, PE/34:0 and PE/34:2. The peaks of m/z 748 and 744 appears highly diagnostic for *Acinetobacter calcoaceticus*.

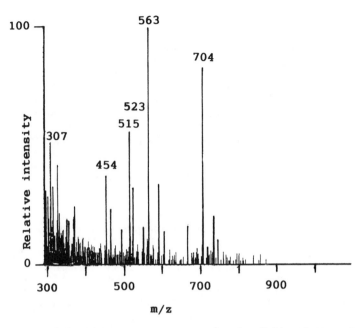

Figure 4. NI-FAB mass spectrum of polar lipids of *Enterobacter cloacae*, adapted from (*31*).

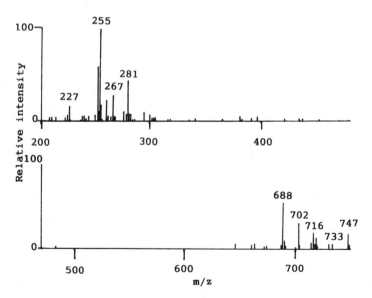

Figure 5. NI-FAB mass spectrum of polar lipids of *Proteus mirabilis*, adapted from (*32*).

Anaerobic Gram-negative Rods.

Prevotella. Two species of *Prevotella* have been examined. *Prev. intermedia* and *Prev. melaninogenica* (Table IV) have distinctive profiles from one another and from other bacteria examined thus far by NI-FAB-MS (Figure 6). For *Prev. intermedia*, the most abundant ions are of m/z 663, 692, 706, 664, 649, and 678 in decreasing order. These correspond to PG/28:1, PE/OH-31:0, PE/OH-32:0, PE/OH-29:0, PE/27:1, PE/OH-30:0. *Prevotella* is known to contain hydroxy fatty acids which would account for the m/z values observed rather than phosphatidylserine of the same nominal m/z. In *Prev. melaninogenica*, the major phospholipid ions are of m/z 589, 590, 587, 603, and 663. These correspond to lipid with fairly short chain fatty acyl substituents rather than peaks in the intermediate range (*vide supra*) resulting from fragmentation. Peaks of m/z 589, 587, 603,and 663 correspond to PG/23:3, PG/23:4, PG/24:3, PG/28:1. The peak of m/z 590 corresponds to PE/25:1.

Capnophilic Gram-negative Rods.

Capnocytophaga. This microorganism prefers low levels of CO_2 for growth. Little is known regarding its polar lipids. However, its major phospholipid peaks are of m/z 574, 588, 662, 575, and 618 (Figure 7). These correspond to PE/24:2, PE/25:2, PE/30:0, PG/22:3, and PE/27:1. One difficulty of interpreting spectra in the absence of other (e.g., chromatographic) data can be seen from the fact that a peak of nominal m/z 662 might conceivably be due to PS/27:1. This illustrates the weakness of peak identifications based upon nominal masses, yet for taxonomic purposes differences between strains are useful data - even in the absence of firm chemical data.

Gram-positive cocci.

Streptococcus. This is a chain-forming coccus that does not produce catalase. *S.mutans* in NI-FAB-MS has major phospholipid peaks of m/z 747, 688, 716, 1077, and 749 (Figure 8). These correspond to anions of PG/34:2, PE/32:1, PE/34:1, unknown, and PG/34:0. The unknown peak appears not to be due to PE, PG, LPG, PI, DGDG, DPG, PS, or acylPG.

Staphylococcus and *Micrococcus.* These catalase-producing non-chaining cocci are similar but differ in oxygen requirements. When examined by NI-FAB-MS, two out of eight most intense peaks are common, viz. 721 and 693 which correspond to PG/32:0 and PG/30:0. In addition, *Staphylococcus* has peaks of m/z 735, 763, 749, and 723 which correspond to PG/33:0, PG/35:0, PG/34:0, and an unknown peak, possibly PG/OH-31:0.

On the other hand, *Micrococcus* has additional peaks of m/z 694, 707, 766, and 768. These correspond to PG/31:0 (m/z 707), and three unknown peaks (*32*).

Non-sporing Gram-positive rods. This term embraces a diverse collection of bacterial genera. In some cases, their classification is poorly understood; some are difficult to identify. Strains of *Corynebacterium xerosis, Mycobacterium smegmatis*, and actinomycetes ('higher bacteria') have been examined by PI-FAB-MS(*28*) and spectra published for *Microbispora parva* and an unspeciated actinomycete (Figure 9) from the Schering collection. *M.parva* was found to contain 'a PE series with fatty acids of increasing chain length' viz. PE/31:0, PE/32:0, PE/33:0, and PE/34:0. Ions of m/z value two integers greater were interpreted as containing a hydroxyl group. Ions due to $[M+Na]^+$ and $[M+K]^+$ were also recorded (*vide supra*). In the author's laboratory a collection of *Lactobacillus* strains have been analyzed by NI-FAB-MS. *L.rhamnosus* has major phospholipid peaks of m/z 761, 762, 759, 787, 733 (Figure 10). These anions correspond to PG/35:1, PE/OH-36:0, PG/35:2, PG/37:2, and PG/33:1. Clearly, they are highly diagnostic for *Lactobacillus*.

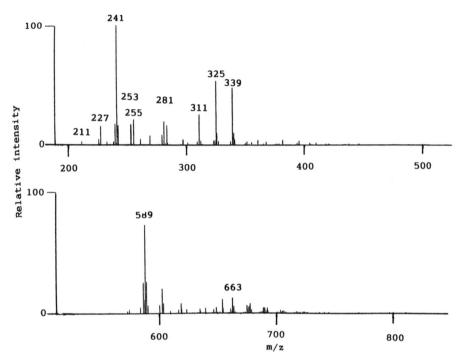

Figure 6. NI-FAB mass spectrum of polar lipids of *Prevotella melaninogenica*.

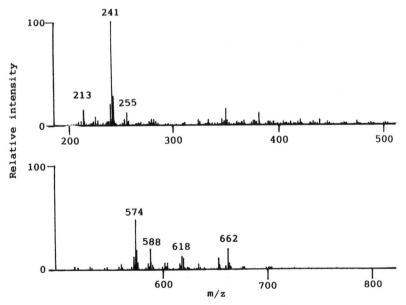

Figure 7. NI-FAB mass spectrum of polar lipids of *Capnocytophaga gingivalis*.

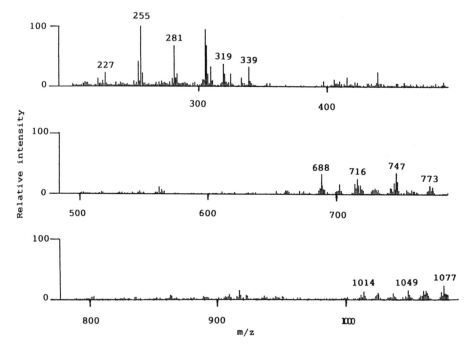

Figure 8.   NI-FAB mass spectrum of polar lipids of *Streptococcus mutans*.

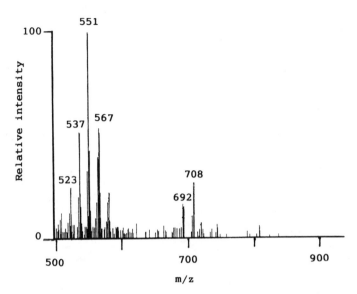

Figure 9.   PI-FAB mass spectrum of polar lipids of an actinomycete, adapted from (28).

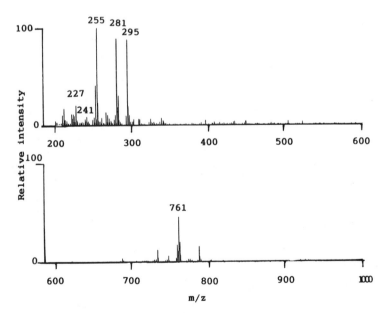

Figure 10.   NI-FAB mass spectrum of polar lipids of *Lactobacillus rhamnosus*.

Sporing Gram-positive rods. One study (*32*) has reported findings for a strain of *Bacillus subtilis*. This mainly has PG followed by smaller contributions due to anions of PE, LPG, and DGDG in NI-FAB-MS spectra. Major peaks are of m/z 721, 693, 722, 694, and 707. These correspond to PG/32:0, PG/30:0, unknown, unknown, and PG/31:0. The unknown peaks are one integer higher than major peaks and would not appear as major peaks if one or two peaks had not been so much more intense than the rest.

Numerical Analysis of Data.

The data presented above have demonstrated that considerable qualitative differences exist in polar lipids of different microbial species and that such differences could have taxonomic use. When more closely related taxonomic groups are investigated, quantitative differences can be sought. Linear regression (*33*) may be employed to assess in a quantitative manner the similarity/dissimilarity of spectra from different microorganisms. Table V presents correlation coefficients obtained by comparing various spectra. A correlation coefficient of unity indicates two spectra are identical. Smaller values of the correlation coefficient indicate greater dissimilarity of spectra. In Table V, the self comparisons are based on the identical spectra. Repeat spectra for the same organism grown and extracted identically have slightly lower values, typically 0.98 or 0.99.

Table V. Comparison of Strains Using the Coefficient of Linear Correlation

Species	A.calcoaceticus	Ent.cloacae
A.calcoaceticus	1.00	0.46
C.freundii	0.48	0.90
Ent.cloacae	0.46	1.00
M.morganii	0.16	0.52
Pr.mirabilis	0.46	0.80
Ser.marcescens	0.19	0.74

Very simply, the analysis for an unknown strain is compared with data in a computer library for a series of known strains. The assumption is made that taxonomic relationships of strains is reflected in differences in their chemical composition. The library strain most similar to the strain being identified is regarded as possibly being the same species (*2,34*). Such techniques (*35,36*) have previously been used for fatty-acid fingerprints obtained by gas chromatography (*2*) which can permit automatic identification based upon chemical data alone. A number of stages are usually followed. Firstly, data are computed to provide proportional values for peaks to facilitate comparison. Secondly, a measure of association is required to compare a chemical analysis for an unidentified strain with that for a 'known' strain in a computer library. This measure can range from a coefficient of correlation, *e.g.*, linear correlation or rank correlation, to a coefficient of similarity of the type used for numerical taxonomy of microorganisms (*34,4,37*).

At this point procedures vary. If a pre-existing classification is used to identify an unknown strain, then a computer can select the highest coefficient obtained with a library strain's data and print out that strain's species name as the identification of the unknown strain, *provided that* a predetermined value for the coefficient for the strain comparison has been attained. This is very important because an unknown strain that is not closely related to any library strain will , nevertheless be more similar (have a higher coefficient with) to one strain in the library than to others. Standard programs used in numerical taxonomy can also print out a nearest neighbour table which lists the several most similar library strains and values of coefficients for strain comparisons of unknown and individual library strains.

The alternative procedure arises when a chemotaxonomy is performed and all the strains are being classified according to their chemical composition. Here, coefficients for all strain comparisons are placed in a matrix of values . The next stage is to 'sort' the values so that similar strains are placed together and unrelated strains are separated. Several different approaches are possible for performing this process which is termed 'cluster analysis'(35,36,37). In the simplest approach, the computer is programmed to search all the values in an unsorted matrix and to select all those values above a set level, *e.g.* strain comparisons where the coefficient attains at least a value of 0.98. Such pairs of strains form the nuclei of growing clusters of like strains which are added to by strains which on re-searching of the matrix are found to have a value of, say, 0.96 with any strain already in a cluster. A new cluster can be formed if the selected value is attained by two strains both of which failed to have a value of 0.98 with any other strain previously. By successively lowering the threshold value and re-examining the matrix, clusters of strains will grow, and new clusters will form. At a sufficiently low value, all clusters will coalesce. The final point is that values clustered at a very high value are equivalent to a species. Clusters formed by aggregation of species clusters represent single genera. When the level of similarity required is dropped to much lower values, some cluster representing genera will be grouped together in very large clusters equivalent to bacterial families. Thus well established published methods can be used to identify and to classify strains on the basis of chemical analytical data alone.

Future Possibilities

The data presented (Table IV) clearly indicate that phospholipid analysis by FAB-MS yields novel information of taxonomic value and not readily gleaned for large numbers of strains by alternative means. Compared with earlier approaches such as GC(2), FAB-MS offers two useful advantages. It is not so susceptible to interlaboratory variation as methods dependent upon GC columns and gas flow and temperature. Secondly, the data collected are more than just a fingerprint. The mass spectra provide information on the chemical identity of the lipids. For example, individual m/z values are related to molecular weights. There is no reason why data collected in laboratories in different continents should not be pooled in multicentre studies, from the chemical view point. However excellent standardization of both microbiological sampling and analytical conditions would be required. Otherwise differences between samples might be due to differing methodology rather than differing microorganism(30). Such studies could provide data both for classification of strains and for identification of individual unknown strains subsequently. Another area of interest to microbiologists will be FAB-MS of polar lipids, and other compounds, as a means of obtaining data on possible virulence factors in disease. As awareness of the technique increases, and especially if FAB-MS becomes a cheaper and more widely available technique so its use in microbiological applications will grow.

Tandem mass spectrometry will make an increasing impact when used in conjunction with FAB for isomeric determination in fatty acids (38).

Similarly useful complementary information will be derived from use of laser desorption MS (39,40,41).

Literature Cited
1.   Goodfellow,M.; Minnikin,D.E. In *Chemical Methods in Bacterial Systematics*; Goodfellow,M.; Minnikin,D.E.,Ed.; Academic Press: London,U.K.,1985,1-15.
2.   Drucker,D.B.*Microbiological Applications of Gas Chromatography*; C.U.P.:Cambridge,U.K.,1981;400-428.
3.   Aluyi,H.S.; Drucker,D.B.*J.Chromatogr*.**1979**,*178*,209-218.
4.   Drucker,D.B.; Jenkins,S.A.*Trans.Biochem.Soc*.**1989**,*17*,245-249.
5.   Tavakkol,A.;Drucker,D.B.;Wilson,J.*Biomed.MassSpectrom*.**1985**,*12*,359-363.
6.   Minnikin,D.E.; O'Donnell,A.G.; Goodfellow,M.; Alderson,G.; Athalye,M.; Schaal,A.; Parlett,J.H.*J.Microbiol.Meth*.**1984**,*2*,233-241.
7.   Batley,M.; Packer,N.H.; Redmond,J.W.*Biochim.Biophys.Acta* **1982**,*710*,400-405.
8.   Tomer,K.B.; Crow,F.W.; Gross,M.L.*Anal.Chem*.**1983**,*55*,1033-1036.
9.   Johnson,R.S.; Her,G.-R.; Grabarek,J.; Hawiger,J.; Reinhold,V.N.*J.Biol.Chem*.**1990**,.8108-8116.
10. Nishihara,M.; Koga,Y.*Biochem.Cell Biol*.**1990**,*68*,91-95.
11. Kramer,J.K.G.; Sauer,J.K.G.; Blackwell,B.B.*Biochem.J*.**1987**,*245*,139-143.
12. Rendell,N.B.; Taylor,G.W.; Somerville,M.; Todd,H.; Wilson,R.; Cole,P.*J.Biochim.Biophys.Acta* **1990**,*1045*,189-193.
13. Heller,D.N.; Cotter,R.J.;Fenselau,C.; Uy,O.M.*Anal. Chem*.**1987**,*59*,2806-2809.
14. Devienne,F.M.; Roustan,J.-C.*C.R.Acad.Sci*.**1976**,*283B*,397-399.
15. Barber,M.; Bordoli,R.S.; Sedgewick,R.D.; Tyler,A.N.*J.C.S.Chem.Comm*.**1981**,325-327.
16. Hollingworth,P.S.;Hollingworth,R.I.;Dazzo,F.B.*J.Biol.Chem*. **1989**,*14*,61-1466.
17. Seid,R.C.; Bone,W.M.; Phillips,L.R.*Anal.Biochem*.**1986**,*155*,168-176.
18. Ballistreri,A.; Garozzo,D.; Giuffrida,M.; Impallomeni,G.; Montaudo,G.*Macromol*.**1989**,*22*,2107-2111.
19. Banoub,J.; Boullanger,P.;Andre,C.;Becchi,M.;Fraisse,D.*Mass Spectrom*.**1990**,*25*,129-130.
20. Gibson,B.W.; Biemann,K.*Biochem.J*.**1984**,*81*,1956-1960.
21. Beckner,C.F.; Caprioli,R.M..*Anal.Biochem*. **1983**,*130*,328-333.
22. Fujii,Y.; Hayashi,M.; Hitotsubashi,S.; Fuke,Y.; Yamanake,H.; Okamoto,K.*J.Bacteriol*.**1991**,*173*,5516-5522.
23. Qureshi,N.; Cotter,R.J.; Takayama,K.*J.Microbiol.Meth*. **1986**,*5*,65-77.
24. Fenwick,G.R.; Eagles,J.; Self,R.*Biomed.Environ.Mass Spectrom*.**1983**,*10*,382-386.
25. Munster,H.;Stein,J.;Budzikiewicz,H.*Biomed.MassSpectrom*.**1986**,*13*,423-427.
26. Hayashi,A.; Mitsuhara,T.; Morita,M.; Kinoshita,T.; Nakamura,T.*J.Biochem*.**1989**,*106*,264-269.
27. Ayanoglu,E.; Wegmann,A.; Pilet,O.; Marbury,G.D.; Hass,J.R.; Djerassi,C.*J.Am.Chem.Soc*.**1984**,*106*,5246-5251.
28. Pramanik,B.N.; Zechman,J.M.; Das,P.R.; Bartner,P.L. *Biomed.Environ.Mass Spectrom*.**1990**,*19*,164-170.
29. Ross,M.M.; Neihof,R.A.; Campana,J.E.*Anal.Chim.Acta* **1986**,*181*,149-157.
30. Aluyi,H.S.; Boote,V.; Drucker,D.B.;Wilson,J.M. *J.Appl.Bact*.**1992**,*72*,80-88.

31.   Aluyi,H.S.;   Boote,V.;   Drucker,D.B.;   Wilson,J.M.;
       Ling,Y.H.*J.Appl.Bact.***1992**,*73*,426-432.
32.   Heller,D.N.;   Murphy,C.M.;   Cotter,R.J.;   Fenselau,C.;
       Uy,O.M.*Anal.Chem.***1988**,*60*,2787-2791.
33.   Platt,J.A.;   Uy,O.M.;   Heller,D.N.;   Cotter,R.J.;
       Fenselau,C.*Anal.Chem.***1988**,*60*,1415-1419.
34.   Drucker,D.B.; Hillier,V.F.; Lee,S.M.*Microbios* **1982**,*35*,139-150.
35.   Sneath,P.H.A.;   Sokal,R.R.*Numerical   Taxonomy* ,Freeman:San
       Francisco,**1973**.
36.   Everitt,B.*Cluster Analysis*,Halsted Press:NY,**1980**.
37.   Jenkins,S.A.; Drucker,D.B.; Hillier,V.F.; Ganguli,L.A.
       Microbios **1992**,*69*,139-154.
38.   Tomer,K.B.; Jensen,N.J.; Gross,M.L.*Anal.Chem.***1986**,*2429-
       2433*.
39.   Seydel,U.; Lindner,B.; Wollenweber,H.-W.; Rietschel,E.T.
       *Eur.J.Biochem.***1984**,*145*,505-509.
40.   Lindner,B.;   Seydel,U.;   Zahringer,U.   In   *Advances   in   Mass
       Spectrometry*;   Todd,J.F.J.;John   Whiley: NY,**1985**,Part   B.
41   Cotter,R.J.; Honovich,J.; Qureshi,N.; Takajama,K.*Biomed.*
       *Environ.Mass Spectrom.***1987**,*14*,591-598.

RECEIVED April 16, 1993

Chapter 4

Microbial Characterization by Phospholipid Profiling

Practice and Problems in Microbial Identification

Mark J. Cole[1] and Christie G. Enke[2]

[1]Drug Metabolism Department, Central Research Division, Pfizer, Inc.,
Groton, CT 06340
[2]Department of Chemistry and Center for Microbial Ecology, Michigan
State University, East Lansing, MI 48823

The variety and relative abundance of membrane phospholipids are
widely thought to be unique for each microorganism. Tandem mass
spectrometry has been shown capable of rapidly and sensitively
characterizing the phospholipid content of crude microbial lipid
extracts. This work confirms that the large data sets containing the
masses and fatty acyl compositions for each phospholipid class
obtainable by this technique include not only valuable taxonomic
information but are also strongly influenced by environmental factors,
including the growth conditions, growth medium, sample history, and
the act of culturing itself. Due to the large size of these data sets and
the limited amount of such data yet available, correlations between the
various elements of the data sets and the factors of identity, history,
activity, and function have not been obtained. Though clearly a
powerful technique, the full utilization of phospholipid profiling by
tandem mass spectrometry for both environment-independent
taxonomy and studies on microbial function and activity awaits the
development of suitable data analysis tools.

A multitude of different biomolecules are needed by each microbial species to perform
the functions required to fill, and survive in, its niche in nature. The natural selective
forces imposed on different microorganisms have required microbes to adapt and
optimize specific sets of biomolecules for survival in specific environments. A few
examples of these biomolecules include the lipids needed for membrane function and
integrity, the respiratory quinones needed for metabolism, and the polysaccharides
needed for protection from the environment. The particular collection of these
molecules incorporated in each microbial species is unique to that species. In fact,
individual subsets of this complete collection of biomolecules are often unique to a
particular species. Identification of these unique subsets can thus provide biomarkers
for identification of microorganisms.

The diversity of lipids in microorganisms signifies a diversity of functions.
Lipids, either directly or indirectly, play a significant role in the specialized functions
each species maintains to survive in its particular environmental habitat. Lipids are
critical to many vital functions including storage, membrane structure and function,
photosynthetic processes, and most energy-generating processes. The types and

0097–6156/94/0541–0036$07.50/0
© 1994 American Chemical Society

distribution of lipids contained in microorganisms can be very species specific (*1-3*). However, not all of the cellular lipids are easily accessed. Phospholipids, which are responsible for the structure of the cell membrane, and thus, ultimately its function, are easily accessed through simple extraction procedures and provide a useful and abundant biomarker for microbial detection and identification. In addition, phospholipids are readily amenable to desorption-mass spectrometric techniques. This makes an attractive package for microbial identification: an abundance of specific information contained within the phospholipid content, coupled with the speed and sensitivity of mass spectrometry.

Despite the above attractions, the use of phospholipids for microbial identification is daunting. Success depends on two factors: achieving high discriminating power and correlating the data with the identification. Mass spectrometry has been shown to provide the needed discriminating power (*4, 5*), and tandem mass spectrometry has been particularly successful (*6, 7*). Only two uses of phospholipid data for the general identification of microorganisms have been reported (*8, 9*), but neither of these have been applied to large data sets.

Paradoxically, the qualities that make membrane phospholipids attractive for microbial identification - the abundance of information contained in the phospholipids and the functional uniqueness of the phospholipid membrane composition for each microorganism - also make analysis of these data difficult. The difficulty arises in two areas: first, the growth conditions, nutritional status, and history of a microorganism can cause changes in its phospholipid profile as the microbe changes its membrane composition in response to its environmental requirements, and secondly, the amount of data generated from the phospholipids rapidly exceeds the capacities of the common data analysis methodologies in existence today.

This research approaches the problem of microorganism identification through: 1) developing a general method for rapidly and sensitively characterizing the phospholipid content and structures in crude lipid extracts; 2) using this methodology to probe the extent to which the phospholipid data varies as microorganisms adjust to environmental stresses; 3) exploring the types of data obtained and the possibilities for new data analysis tools.

Phospholipid Profiling

The phospholipids with which this research is concerned are the glycerophospholipids (Structure 1). Glycerophospholipids consist of four primary functional groups: a glycerol-3-phosphate core on which two fatty acyls (R, R') have been esterified to the two free hydroxyl groups in the *sn*-1 and *sn*-2 positions, and a second alcohol (Y) has been esterified to the phosphate group in the *sn*-3 position. The exception to this basic structure is phosphatidic acid, which only contains the phosphate group. The head group (Y) is the functional group that defines the specific class to which the phospholipid belongs, while the fatty acyls distinguish the individual phospholipid molecular species within each class. Examples of head groups from several phospholipid classes are shown in Figure 1. Crude lipid extracts are obtained using a modified Bligh-Dyer procedure (*10*).

Phospholipid Dissociation Under Low-Energy CAD Conditions.

When phospholipid ions undergo low-energy CAD, only a few different dissociation product ion masses are formed. However, the CAD product ions that are produced are the result of cleavages at specific points that are common to all classes of phospholipids. These cleavages occur around the phospholipid functional groups and thus provide significant structural information about the phospholipid. The low-energy CAD fragmentation of phospholipids for both positive and negative ions is shown in Figure 2. In the positive ion mode, the major reaction occurring is cleavage of the phosphate/glycerol bond resulting in the loss of the polar head group as a

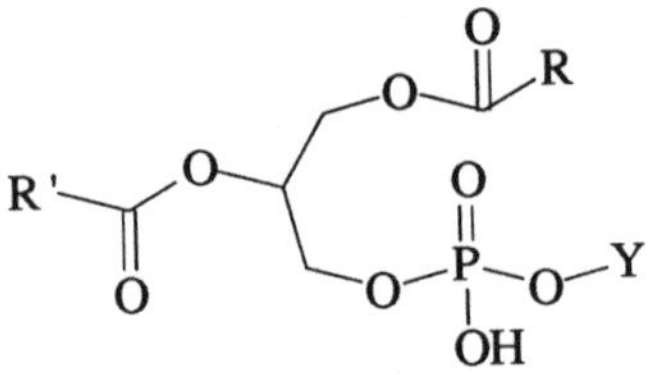

1

PHOSPHOLIPID	HEAD GROUP (Y)

Phosphatidylethanolamine
(PE)

Phosphatidylglycerol
(PG)

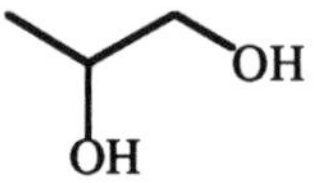

Phosphatidylcholine
(PC)

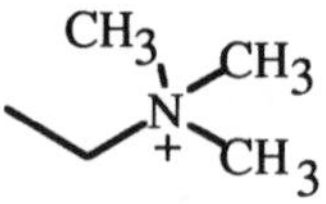

Phosphatidylserine
(PS)

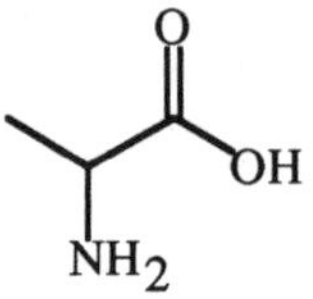

Phosphatidylinositol
(PI)

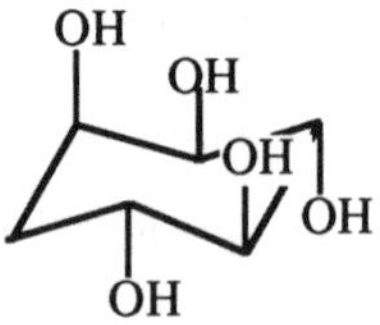

Figure 1. Examples of the head groups from five phospholipid classes.

neutral, while the rest of the ion retains the charge. In the negative ion mode, one or the other fatty acyl is cleaved and retains the charge, while the rest of the ion is lost as a neutral. This research has taken advantage of the fact that phospholipids always cleave around their functional groups to develop a general scheme for the rapid characterization of phospholipid classes present in microorganisms, including information on the fatty acyls present on the species within each class. This scheme is a sequence of positive ion neutral loss and precursor scans, and negative ion CAD product scans obtained with a triple quadrupole mass spectrometer.

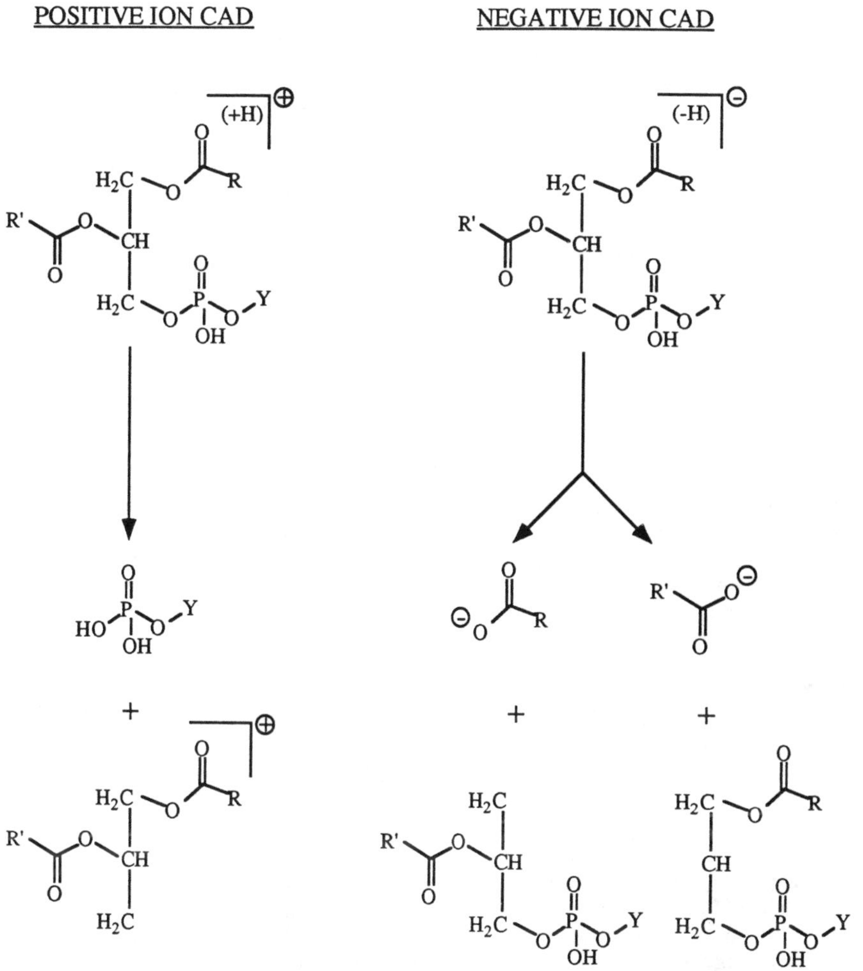

POSITIVE ION CAD   NEGATIVE ION CAD

Figure 2. Low-energy CAD fragmentation of phospholipids for positive and negative ions. (Adapted from ref. 7.)

Phospholipid Data Space. Phospholipid data are obtained by this technique on two levels. The first level is referred to as the *(class) mass profiles* which are the separate mass spectra for each phospholipid class. From these data, information is obtained regarding the phospholipid classes present, the masses of individual phospholipid species within each class, and the relative intensities of the species within each class. The second data level consists of the dissociation product spectra of individual phospholipid species and is referred to as *(class)(fatty acyl) formula data*. These data can be interpreted to provide empirical formula information on the fatty acyls and structural information on each phospholipid species. The data levels and their interrelation are shown as a data space diagram in Figure 3. In this space, each phospholipid class forms its own data plane with the mass profile of its species contained along one dimension of the plane and the fatty acyl masses for each of its species along the other dimension. Low-energy CAD does not yet allow the complete structure of the fatty acyl groups to be obtained.

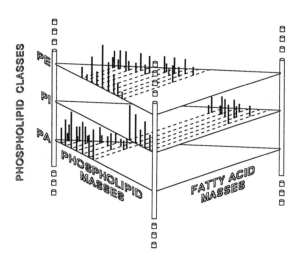

Figure 3. The FAB/MS/MS data space obtained by this technique. The masses of the phospholipids in each class are obtained by neutral loss or precusor scans specific for each class. The CAD product spectrum of each phospholipid species provides the masses and positions of the fatty acyls. (Reproduced from ref. 7. Copyright 1991 AmericanChemical Society.)

General Profiling Scheme. The first step in the general scheme is to obtain the positive ion neutral loss scans for loss of the masses of the different phospholipid polar head groups. The instrumental parameters used in these experiments have been previously published (7). The instrument is programmed to scan a series of possible polar head group neutral losses in the positive mode. No CAD gas is added to the collision chamber other than the residual gas remaining in the analyzer manifold. This produces the optimum conditions, and the dissociation observed may be predominantly the result of metastable decomposition. *(Class) mass profiles* are obtained, of all the phospholipid classes scanned, that approximate the relative abundances of all species within each phospholipid class. The resulting spectra are free of any non-phospholipid peaks, and many peaks of interest which might have been hidden previously in the chemical background of the conventional mass spectrum show up clearly. An example of the use of neutral loss scans for collecting *(class) mass profiles* of two of the phospholipid classes contained in *Proteus vulgaris* is shown in Figure 4. Note that the peaks due to phosphatidylglycerol (neutral loss 172) are completely buried in the chemical background of the conventional spectrum. This added detectability is a direct result of the additional selectivity afforded through tandem mass analyzers.

The relative intensities of the phospholipid species obtained in each *(class) mass profile* reflects the relative intensities of the species in the conventional mass spectrum. Apparent differences in intensities between the conventional spectrum and the neutral loss spectra are seen when more than one phospholipid class contributes to the abundance of a particular peak in the conventional mass spectrum. These overlying peaks are resolved by the neutral loss spectra. The only exception that was found to using neutral loss scans is in the characterization of phosphatidylcholine. Phosphatidylcholine fragments so that the head group usually retains the charge. While the neutral loss scan does work for phosphatidylcholine, greater sensitivity was obtained through the use of a precursor scan for the head group ion. Precursor spectra are obtained at 30 eV collision energy using argon as a CAD gas at a pressure of 0.5 mtorr. A precursor scan for m/z 184 produces the *(class) mass profile* for phosphatidylcholine.

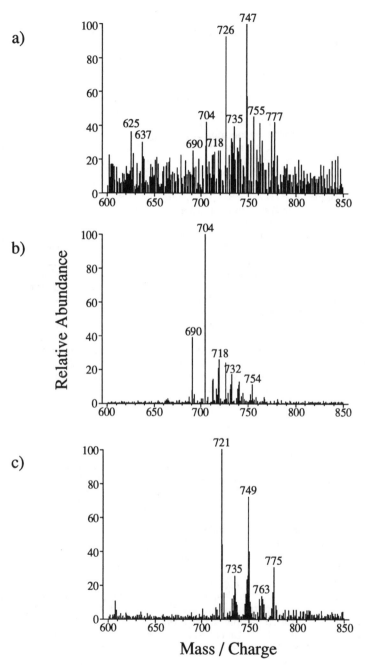

Figure 4. (a) Conventional mass spectrum of a *Proteus vulgaris* extract. (b) Neutral loss spectrum of 141 u; specific for phosphatidylethanolamine. (c) Neutral loss spectrum of 172 u; specific for phosphatidylglycerol. (Reproduced from ref. 7. Copyright 1991 American Chemical Society.)

The two remaining pieces of information needed for a complete phospholipid profile are the composition, structure, and relative positions of the fatty acyls on the glycerophosphate core. Much of this information can be directly obtained from the negative ion CAD product spectra of the individual phospholipid species detected in the series of *(class) mass profiles* obtained from the neutral loss scans previously performed. These spectra are obtained under the same CAD conditions as the precursor spectra. As shown in Figure 2 for [M-H]⁻ ion dissociation, the two major fragments present in a negative ion CAD product spectrum are the peaks due to the two fatty acyls contained on the phospholipid ion. The mass/charge value of the fragment ion corresponds to the molecular weight of the free fatty acyl, less one hydrogen. In this manner, the number of carbons contained in the fatty acyl may be determined, along with the degree of unsaturation. It is important to note that from these spectra, only the empirical formulae of the fatty acyls may be determined with certainty. At the present time, low-energy CAD product spectra cannot determine the location of double bonds on a fatty acyl, differentiate between a point of unsaturation and a cyclic (cyclopropyl) structure, or differentiate normal from branched fatty acyls.

The negative ion CAD product spectra also can provide information on the relative positions of the two fatty acyls on the phospholipid. Zirrolli, *et al.* (*11*) have postulated preferential formation of the carboxylate anion from the fatty acyl at the sn-2 position (closest to the head group) over that of the fatty acyl at the sn-1 position. When this is true, the CAD product ion peak with the greater intensity is due to the fatty acyl from the sn-2 position. While this general rule apparently does not apply to phospholipids containing highly unsaturated fatty acyls or when large differences in the chain lengths of the two fatty acyls occur (*12*), it does hold for phospholipids containing fatty acyls with fewer than three points of unsaturation and with chain lengths differing less than 10 carbons from each other. Because fatty acyls which fit the exceptions to this rule are extremely rare in bacterial phospholipids, this rule is being used to obtain positional information about the fatty acyls present in our samples. The exceptional fatty acyls are easily diagnosed and the data treated accordingly. Figure 5 is an example of the use of a negative CAD product spectrum for *Escherichia coli*. The peak at m/z 702 represents the [M-H]⁻ ion of a phosphatidylethanolamine species. The two CAD product ion peaks obtained at m/z 255 and m/z 267 correspond to a C16:0 fatty acyl and a C17:cyc fatty acyl, respectively. As shown by the difference in peak intensities, the C17:cyc fatty acyl occurs at the sn-2 position on the phospholipid.

An automated instrument control procedure has been written that performs the necessary neutral loss or precursor scans, in the positive ion mode, for all 9 phospholipid classes in which this research is interested, switches the instrument into the negative ion CAD product mode, and collects the individual CAD product spectra from the 10 major peaks present in each of the phospholipid classes previously detected. The resulting data file contains up to the 9 possible *(class) mass profiles* and the CAD product spectra of up to 90 phospholipid species. The automated procedure can collect this information from a sample in under ten minutes. Only a single loading of the sample probe is required for samples containing the extracted lipids from the equivalent of a single colony (approximately 10^6 cells). The ability to automate, rapidly switch between scan modes, and perform multiple experiments in a single analysis becomes very important when a large quantity of data needs to be collected from a potentially limited sample. The efficiency of the triple quadrupole mass spectrometer in fulfilling these functions plays a key role in obtaining the microbial phospholipid data described above.

Sensitivity. The increase in selectivity due to the use of tandem mass spectrometry gives outstanding detection limits that are unattainable for these samples in conventional mass spectrometry. The top spectrum in Figure 6 is a neutral loss scan for phosphatidylethanolamine (neutral loss of 141 u) of one picogram of a mixture of

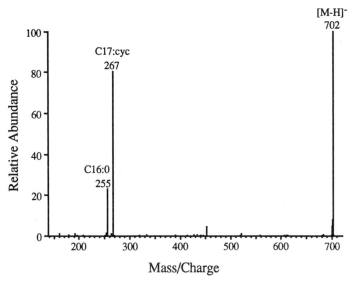

Figure 5. Negative-ion CAD product spectrum of a phosphatidyl-ethanolamine species at *m/z* 702 showing the two fatty acyl ions and their relative intensities (C16:0 is a 16-carbon fatty acyl with no unsaturation; C17:cyc is a 17-carbon propyl fatty acyl). These assignments were made from correlating molecular weights with known fatty acyl structures. From these correlations, *m/z* 255 most likely represents C16:0, and *m/z* 267 most likely represents C17:cyc or C17:1. Since the occurrence of C17:1 is somewhat rare in bacteria, 267 most likely represents C17:cyc.

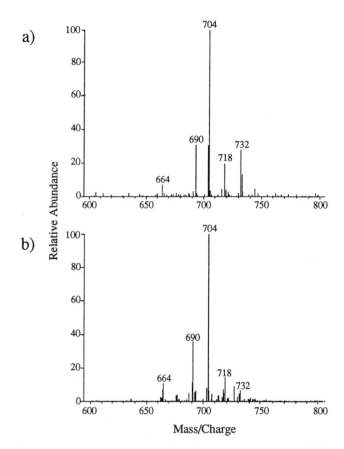

Figure 6. (a) Phosphatidylethanolamine mass profile of 1 pg of PE extracted from *E. coli*; equivalent to 10 cells. (b) PE mass profile of 1 μg of PE extracted from *E. coli*; equivalent to 10^7 cells. (Reproduced from ref. 7. Copyright 1991 AmericanChemical Society.)

phospholipids that are mostly phosphatidylethanolamine species extracted from *E. coli*. This is roughly the phospholipid content of ten *E. coli* cells. For comparison, the bottom spectrum shows the same scan of 1-microgram of the same mixture; equivalent to 10^7 cells. All of the phosphatidylethanolamine species present in the 1-microgram mixture have been detected in the one picogram spectrum. This extremely low sample size requirement allows for the analysis of phospholipids at bacterial levels found in natural samples without prior cultivation of the organisms. Most techniques require separation of microorganisms from each other and amplification of the selected microbe. The methodology described here can provide enough sensitivity so that cultivation may not be required for amplification, but the sample must be still be organized in such a way as to contain a single species or only a few species. This would require some means of separation.

Environmental Effects on the Phospholipid Content of Microorganisms

The ultimate goal of microbial identification is the rapid, unerring, taxonomic identification of a single microbe without prior cultivation from its natural environment. Phospholipid profiling using tandem mass spectrometry begins to provide the analytical capabilities needed to reach this goal: a large abundance of information can be obtained from a relatively small number of microbes in just a few minutes. However, the data analysis aspect of the identification requires that the microbes under investigation maintain unique and rather constant membrane phospholipid contents. Since environmental control must be sacrificed to accomplish the goal of direct identification of microorganisms, the degree to which microorganisms vary their phospholipid content in response to environmental changes must be determined before phospholipid profiling techniques can achieve their full potential.

Effects of Growth Temperature and Time on Phospholipid Profiles. The first investigation into the variability of phospholipid content due to the microorganism's environment was the study of the effects of growth temperature and growth time on the bacteria *Escherichia coli* and *Bacillus subtilis*. For the temperature study, the microorganisms were grown on trypticase soy agar for 24 hours at 23, 30, and 37 C, after which they were harvested and their phospholipid profiles obtained. Changing the growth temperature had a profound effect on the relative abundances of the phospholipids in *B. subtilis*, as shown in Figure 7. In general, the bacteria preferentially produced shorter chain-length fatty acyls at the cooler temperatures and longer chain-length fatty acyls at the warmer temperatures. As the temperature progresses from 23 to 37 C, the smaller molecular weight lipids (which represent the shorter chain-length fatty acyls) decrease in relative abundance while the larger molecular weight lipids increase in relative abundance. Similar data were obtained for the experiment with *E. coli*. These changes are not unexpected. Bacteria need to maintain a certain degree of membrane fluidity, and accomplish this by adjusting the chain-lengths of the phospholipid fatty acyls. At lower temperatures, an abundance of shorter-chain fatty acyls allow a more fluid membrane, while at higher temperatures, the longer-chain fatty acyls decrease the membrane fluidity.

The effect of growth time at various growth temperatures for *E. coli* is shown in Figure 8. The phospholipid profiles obtained changed considerably at 24 C, but only minor changes were observed at 37 C. Similar results were obtained for the experiment with *B. subtilis*. These changes reflect the growth stage of the organism. As the microbe grows, the fatty acyl chain-lengths are gradually increased until the optimum fatty acyl combination is obtained, after which the organism enters a stationary phase in its growth. At the cooler temperatures, the organism takes longer to reach this stationary phase. Since 37 C is the optimum growth temperature for *E. coli* and *B. subtilis*, they rapidly (within 24 hours) reach their stationary phases at this

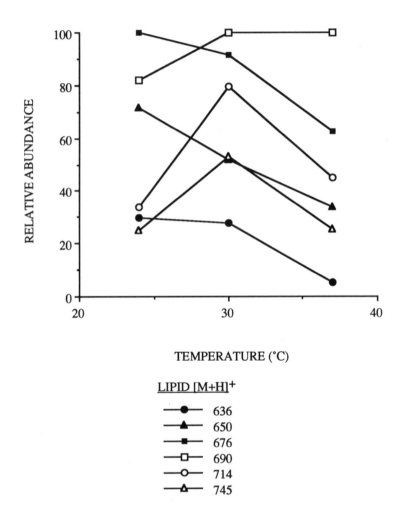

Figure 7. Phospholipid abundance as a function of growth temperature for *Bacillus subtilis* after 24 hours growth time.

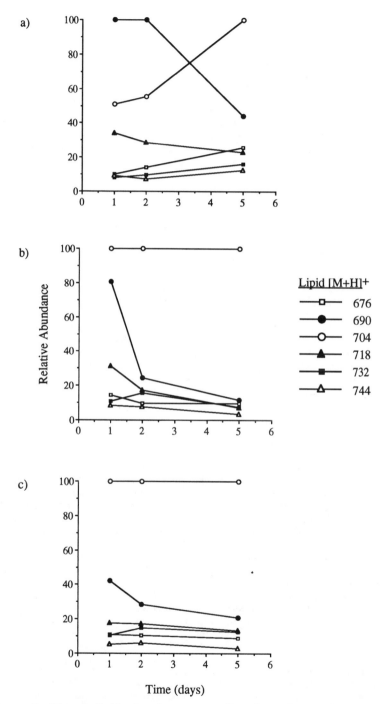

Figure 8. Phospholipid abundance as a function of growth time for *Escherichia coli* at: (a) 24 C; (b) 30 C; and (c) 37 C.

temperature. This is reflected by the fact that only minor changes occur at this temperature. When the microorganisms are grown at their optimal temperatures, growth time has little effect on the phospholipid profiles obtained.

Throughout these studies, the only changes occurring in the membrane phospholipids were in the chain-lengths of the fatty acyls attached to the phospholipids. No new classes of phospholipids were observed to appear as the growth conditions changed, nor were any classes of phospholipids observed to disappear. The effects on microbial fatty acid content due to changes in growth time and temperature have previously been noted (1,13). However, the phospholipid measurements provide an important new insight: no classes of phospholipids are changed during changes in growth times and temperatures, the only changes occur on the fatty acyls contained within each phospholipid class. This evidence strongly suggests that bacteria conserve the types of phospholipids they retain in their membranes, and provides important information for developing data analysis techniques.

Effects of Growth Media on Phospholipid Profiles. During the course of an investigation into the differentiation of food-borne contaminants, the observation was made that the application of a nutritional stress to a microorganism could cause a change in the membrane phospholipid profile. The goal of the study was to differentiate *Salmonella* species from *E. coli* and *Citrobacter freundii*. *Salmonella* species are very difficult to correctly differentiate from *Escherichia* species using common microbial techniques. *Escherichia* and *Salmonella* have up to 50 percent of their DNA sequences in common (14). In the food testing industry, this presents a formidable problem. The methodology currently used to differentiate *Salmonella* from *Escherichia* involves complex immunological assays that can take up to 7 days for results. The cost of storage and spoilage incurred during this testing time can be very expensive. Also, the implications of time and expense in discovering a *Salmonella* contamination 7 days downstream of a production process could be enormous. The food industry is very interested in shortening the time period needed to obtain results.

The phosphatidylethanolamine mass profiles obtained from *S. abaetetuba*, *E. coli*, and *C. freundii*, grown on trypticase soy agar at 37 C for 48 hours, are shown in Figure 9. From these spectra, *C. freundii* is easily differentiated from the other two bacteria, but *E. coli* and *S. abaetetuba* are nearly identical. The similarities between *E. coli* and *S. abaetetuba*, and the differences of these two organisms from *C. freundii*, were conserved throughout the complete phospholipid profiles of all three organisms (space limitations prevent the illustration of the complete phospholipid profiles here). However, all three species were readily differentiated by their phospholipid profiles when grown in tetrathionate broth at 37 C for 48 hours. Tetrathionate broth is partially selective for *Salmonella*. Other bacteria, such as *E. coli* and *C. freundii*, also can grow in this medium, but they do not grow as easily or as fast as *Salmonella*. The phosphatidylethanolamine mass profiles obtained are shown in Figure 10. Differences among the organisms continue throughout the total phospholipid data space, which provides a unique and characteristic information set for each microorganism. Again, for brevity, only enough information to make this point is presented in the illustration. As with the previous growth conditions study, the only changes occurring in the membrane phospholipids were in the relative abundances of the fatty acyls attached to the phospholipids. No new classes of phospholipids were observed to appear as a result of a change in nutrient media, nor were any classes of phospholipids observed to disappear. This is further evidence for the importance of conserving the types of phospholipid classes contained in the membrane.

These results illustrate the importance of the role the growth media can play. Tetrathionate broth puts a selective pressure on *E. coli* and *C. freundii*, while *Salmonella* grows freely. As a result, little difference is observed between the phospholipid profile of *Salmonella* grown on trypticase soy agar and that of

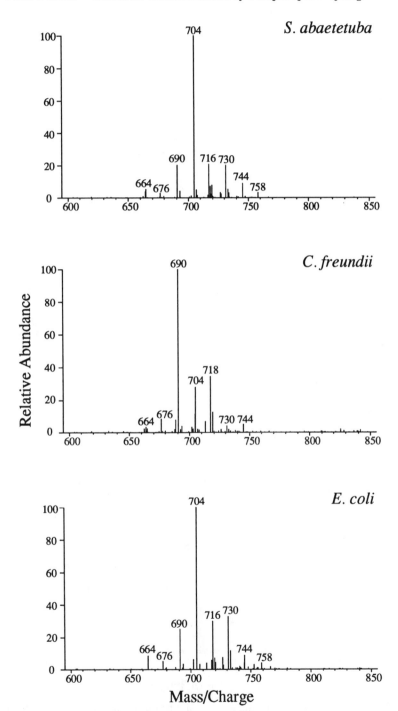

Figure 9. Phosphatidylethanolamine mass profiles from three bacterial food-borne contaminants grown on trypticase soy agar.

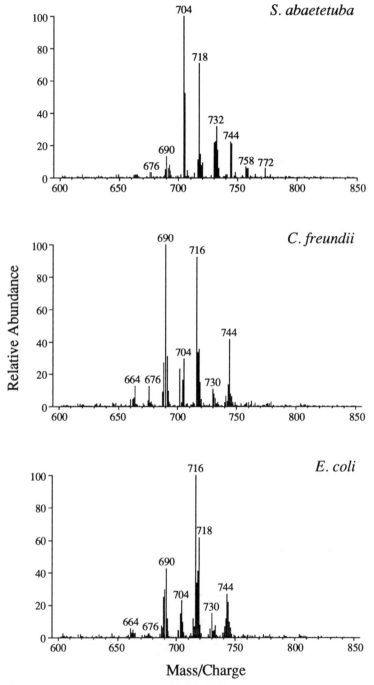

Figure 10. Phosphatidylethanolamine mass profiles from three bacterial food-borne contaminants grown in tetrathionate broth.

Salmonella grown in tetrathionate broth. On the other hand, large differences are observed in the phospholipid profiles of *E. coli* and *C. freundii* grown on the two media. The important observation to be made is that differentiation of closely-related microorganisms can be forced through the application of an environmental stress to one or more of the organisms. For a laboratory wishing to differentiate closely-related organisms, such as food-borne contaminants, application of an environmental stress could be a useful technique. However, control over a microorganism's nutritional environment must be relinquished if the goal of direct identification from raw samples is to be achieved.

Effects of Sample History on Phospholipid Profiles. An interesting observation was made while studying infections of the amoeba *Naegleria* by *Legionella pnuemophila* (*15*). In these experiments, phospholipid profiles were obtained from unexposed amoebas, amoebas infected with the *Legionella* organism, and amoebas that were infected, then subsequently cured by gentamycin. The phospholipid profiles from the unexposed and cured amoebas were expected to be identical, but several distinct differences were readily apparent. The phosphatidylcholine mass profiles obtained from unexposed and cured amoebas are shown in Figure 11. The relative abundances of a few of the fatty acyls contained on phosphatidylcholine change considerably between the two samples. Most notable: a fatty acyl contained on m/z 758 in the unexposed sample saturates to form m/z 760 in the cured sample, a fatty acyl contained on m/z 732 unsaturates to form m/z 730 in the cured sample, and the relative abundances of m/z 770, 744, and 786 change dramatically. The differences in the phospholipid profiles between the unexposed and cured samples occurred only in the phosphatidylcholine portion of the data space; however, phosphatidylcholine was the major contributor to the phospholipid content. In addition to the differences noted above, the unexposed sample contained the glycolipid monogalactosyldiacylglycerol, while the cured sample did not. Diacylglycolipids are similar in structure to phospholipids, only containing a sugar moiety as a head group instead of a phosphate/alcohol moiety. Since the sugar is lost as a neutral during CAD, the method used for profiling phospholipids also works for diacylglycolipids.

These data suggest that the *history* of a microbial sample can have a dramatic effect on the chemical composition of the organism. In particular, this is the first indication of a complete disappearance of a whole class of a chemical biomarker. This has profound implications for data analysis because, as with the nutritional environment, control over the history of a sample must be sacrificed to achieve the goal of direct identification of microorganisms.

The Effects of Culturing on Phospholipid Profiles. The observations made in the above experiments beg the question of whether an organism should be cultured as a pretreatment before collecting data on that organism. Ignoring the fact that culturing moves us further away from our goal of direct identification, can the act of culturing an organism change the organism? Consider *Borrelia burgdorferi*, the causative agent of Lyme disease and a very difficult bacteria to culture outside of its host. The wild organism, obtained from chipmunk ear punches, starts out very sickly in the laboratory and does not grow well initially. However, after several passes through culturing, it begins to grow easily and soon becomes readily culturable in large numbers. The only problem is that the organism present in these cultures is no longer pathogenic when re-introduced into a host. The act of culturing appears to have changed the function of the organism. Is this change reflected in the profiles of phospholipids or other biomarkers? At this time, the question remains unanswered. However, tandem mass spectrometry would be the ideal analysis technique for this

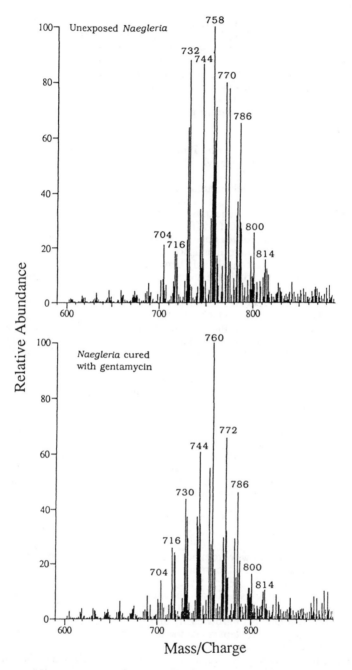

Figure 11. Phosphatidylcholine mass profiles from *Naegleria* amoebae. The top spectrum is from *Naegleria* that have not been exposed to *Legionella pnuemophila*; the bottom spectrum is from *Naegleria* that were previously infected with *Legionella* and subsequently cured by gentamycin.

investigation. The limitation of a few hundred cells obtainable for a wild *Borrelia* sample requires the high sensitivity and selectivity obtainable with tandem mass spectrometry.

Issues Involved With Data Analysis

The investigations described above show evidence that bacteria may change their membrane phospholipid contents as they react to environmental stresses, such as changes in growth temperature, growth time, and growth media. In addition, the phospholipid content of a microorganism can even be dependent on that microbe's past history; and the very act of culturing a microorganism may change that organism. The variability in phospholipid profiles can range from as minor a change as a greater or lesser abundance of a few fatty acyls, to as profound a change as the appearance or disappearance of an entire class of phospholipids. As previously mentioned, if the goal of direct identification of microorganisms without prior cultivation is to be achieved, much of the control over a sample's environment and history will have to be relinquished. The inevitable variability that must remain unaccounted puts stringent demands on the data collection and analysis. Any quantitative differences occurring among spectra from the same species must be less than the interspecies variation for identification to be successful.

Potential Information Contained in a Phospholipid Profile. Not all of the information contained in a phospholipid profile is highly susceptible to environmental changes. For example, the types of phospholipid classes present, their relative abundances, and the presence of particular species of phospholipids were highly conserved among the bacteria studied in the food-contaminant experiments. In addition, much of the phospholipid profile of the gentamycin-cured *Naegleria* did not change from that of the unexposed organism. It is possible that the types of changes a microorganism makes in its membrane in order to interact with its environment are consistent and well-defined. In this light, the phospholipid profile data can be viewed as potentially including: 1) data that are useful for the taxonomic identification of an organism, and 2) data that correspond to a particular quality, function, or environmental state of an organism. The first type of data would be stable and unique to a specific microorganism and would allow its identification. The second type of data would be stable for a particular quality or function and, thus, would be unique to a group of microorganisms. Since the same species might or might not possess a particular microbial quality or function, and certain qualities and functions can be found in different species and/or be transferred among species, these data would more broadly correlate organisms. If these two types of data were first separated within the complete phospholipid profile, the two subsets obtained would become more useful for their particular data analysis functions.

An example of data that are useful for identification of a particular microorganism can be found in the phospholipid profiles of *Legionella* bacteria. In these organisms, long-chain (>20 carbon) fatty acyls occur only on phosphatidylcholine and phosphatidyldimethylethanolamine, and not on any of the other classes of phospholipid or glycolipid contained in these organisms. This information is fairly unique to *Legionella* bacteria and allows for their differentiation from other bacterial species. The information is also unique enough within the *Legionella* to allow identification of individual species. An example of data corresponding to a particular microbial function can be found in thermophilic bacteria. These bacteria thrive on, and even require, very high temperatures for growth and survival. Maintaining membrane integrity becomes a problem at high temperatures for most bacteria because the increased entropic forces cause liquefaction and subsequent degradation of their membranes. Thermophilic bacteria overcome this problem through a unique 30-carbon dicarboxyl fatty acyl (*16*), which is illustrated in Structure

2. The length of this fatty acyl is the exact length required to transverse the membrane and place the free carboxyl group on the opposite, hydrophilic, side of the membrane, as shown in Figure 12. This action helps to hold the membrane together in much the same way that reinforcement rod strengthens a concrete pillar.

Data Analysis Techniques. The sensitivity of mass spectrometry, from the standpoint of the uniqueness of specific phospholipid patterns to individual microorganisms, should lead to its widespread use for microbial characterization and identification. In fact, this has not been the case. While several researchers have used mass spectrometry to analyze phospholipid species from microorganisms (*4-7*), only two uses of phospholipids for the general identification of bacteria have been reported (*8,9*). The data analysis techniques used in these methods include visual correlation of spectra, library searching, and cluster analysis.

Visual Inspection of Data. On a limited scale, comparison and matching of spectra among a small number of microorganisms can be done using the human brain for pattern recognition. However, this visual correlation is based on an acquired familiarity with the similarities and differences that occur among the patterns that characterize different microorganisms. An appreciation of which differences constitute the normal variance for a single organism, and which are valuable in distinguishing organisms, must be obtained through the observation of many spectra. This method rapidly fails for all but the smallest and most well-behaved data sets. In addition, only a few dimensions of data can be correlated; most of the data must be ignored. Given the large number of distinguishing parameters available in the complete phospholipid profiles, the need for computerized techniques becomes apparent.

Library Searching. Statistical techniques and spectral matching using library searches can accommodate larger data sets than visual correlation. For non-variable data, these methods can be very good, and mixtures can be deconvoluted. In addition, a statistical correlation can provide some indication of the degree of similarity or differences among data sets or even portions of data sets. However, these techniques can have very poor resolution when the data sets contain even small variations, they do not perform well when an unknown spectrum is encountered that is not contained in the library, and they can not accommodate multi-dimensional data on the scale of a complete phospholipid profile. Also, library searching does not provide a means for separating data that are useful for taxonomic identification from those that correspond to a particular quality of the microorganism.

Cluster Analysis. Cluster analysis begins to address the issue of separating the types of data found in a phospholipid profile by providing a visual representation and quantitative index, in the form of a dendrogram, of the usefulness of a particular dimension of data for an identification. The food-borne contaminant experiments described earlier were expanded to include the use of cluster analysis. Software to perform this type of analysis was developed in-house by Mr. Eric Hemenway to run on an IBM-AT computer. In this investigation, a second species of *Salmonella* (*S. enteritidis*) was included to provide an even more difficult identification problem, along with *Listeria monocytogenes*, another contaminant of great concern to the food industry. Replicate (N=3 or 4) spectra were obtained for each organism to determine the sensitivity of the method for variances in the intraspecies data sets. The data sets consist of individual *(class) mass profiles*, which provide only the species of phospholipids present within each class of phospholipid. The remainder of the phospholipid profiles, containing the information on fatty acyl types and positions, can not be accommodated by this technique at the present time.

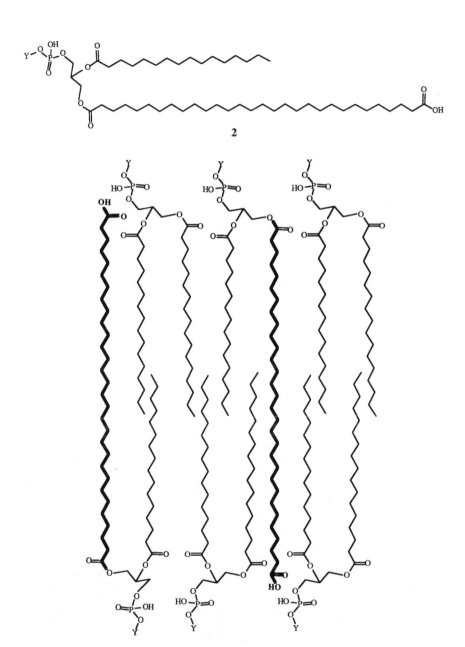

2

Figure 12. Illustration of the incorporation of 30-carbon dicarboxyl fatty acyls into the membrane of a thermophilic bacteria. These fatty acyls transverse the membrane and are anchored on the opposite hydrophilic side by the terminal carboxyl group.

The dendrogram obtained from the cluster analysis of the phosphatidylethanolamine mass profiles is shown in Figure 13. In a dendrogram, the similarity index of two organisms (or sets, or groups of organisms) is given by the position of the vertical lines connecting the elements to be compared. The higher the similarity index, the more data features within the particular phospholipid class they share in common, and the more closely related are the spectra. Phosphatidylethanolamine provides a high degree of discrimination of the *Salmonella* species from *E. coli* and *C. freundii*, but only a moderate discrimination between *E. coli* and *C. freundii*, and almost no discrimination between the two *Salmonella* species. *Listeria monocytogenes* did not contain phosphatidylethanolamine, therefore its differentiation from the rest of the contaminants is completed at this level. The dendrograms constructed from the remaining phospholipid mass profiles are shown in Figure 14. Phosphatidylglycerol provides a degree of discrimination among the organisms similar to that of phosphatidylethanolamine, and phosphatidic acid provides a moderate to good degree of discrimination among all of the bacteria studied. The most useful phospholipid class for discriminating between the two *Salmonella* species is phosphatidyldimethylethanolamine. The profiles from this phospholipid share almost no similarities between the two *Salmonella* species and between *E. coli* and *C. freundii*, thus allowing their unambiguous differentiation. However, these profiles provided no discrimination between *S. abaetetuba* and *E. coli*, and only moderate discrimination between *S. enteritidis* and *C. freundii*. Phosphatidylmonomethylethanolamine discriminates among *E. coli*, *C. freundii*, *S. abaetetuba*, and *L. monocytogenes* (which does not contain this phospholipid). Unfortunately, data were not collected for *S. enteritidis* due to a sampling error, so the degree to which PMME discriminates this organism from the others remains unknown.

Cluster analysis indiscriminately groups microorganisms based on shared phospholipid data. This provides the first means for discriminating between data that correlate particular qualities of the microorganisms and data that are taxonomically unique to each microorganism. The importance of viewing the data in this manner is underscored by the results obtained when cluster analysis is applied to a combination of all the phospholipid mass profiles. When the data from all of the phospholipid mass profiles, each of which clearly discriminates among two or more of the organisms by themselves, were combined in a cluster analysis, the resulting dendrogram did not provide the expected differentiation of all four species. Although *L. monocytogenes* was unequivocally differentiated based on the absence of whole phospholipid classes, the distinction among the other organisms seemed to be "blurred" by data not taxonomically unique to the organisms. The non-descript data for the *Salmonella* species interfered with the clearly discriminating data, and a similar problem occurred for *E. coli* and *C. freundii*.

Cluster analysis is only a small step in the right direction for analyzing these types of data. Although it provides a means for determining the usefulness of data for specific purposes, it does so in only a limited fashion. Restrictions include: 1) due to software limitations, this technique only works on a relatively small number of data features and the degree of filtering required to reduce the amount of data used for analysis may eliminate data that are useful for discrimination; 2) the results from each individual phospholipid mass profile must be manually evaluated for each set of organisms; 3) as with the other data analysis techniques, cluster analysis can not accommodate the amount of data contained in a complete MS/MS phospholipid profile, and as a result, data that may provide a higher level of discrimination are not considered in the analysis.

Unexplored/Underdeveloped Territory. Techniques for collecting data from microorganisms have been well-developed. Mass spectrometry has provided, with remarkable sensitivity, information describing numerous chemical biomarkers

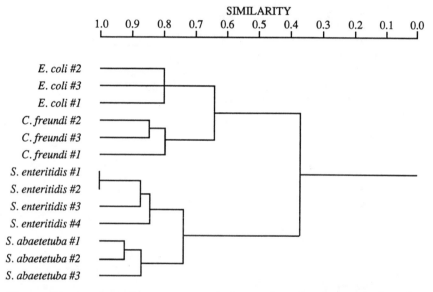

Figure 13. Dendrogram constructed from the cluster analysis of phosphatidylethanolamine mass profiles obtained from several bacterial food-borne contaminants.

PHOSPHATIDYLGLYCEROL (PG)

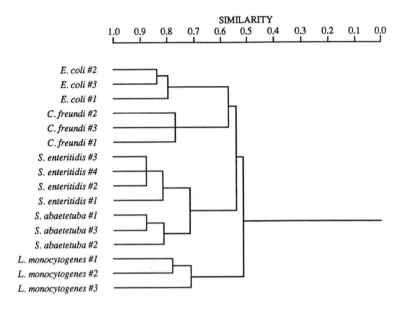

PHOSPHATIDIC ACID (PA)

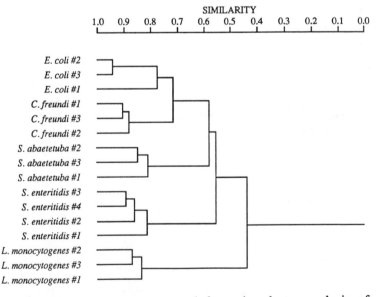

Figure 14. Dendrograms constructed from the cluster analysis of the remaining phospholipid mass profiles obtained from the bacterial food-borne contaminants.

PHOSPHATIDYLDIMETHYLETHANOLAMINE (PDMA)

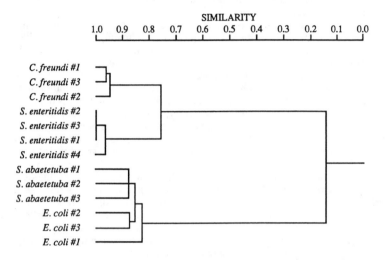

PHOSPHATIDYLMONOMETHYLETHANOLAMINE (PMME)

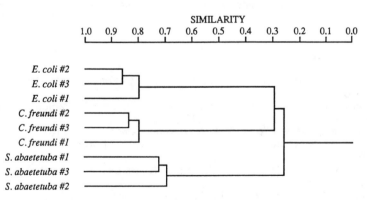

Figure 14. Continued.

contained in microorganisms, including the phospholipids, fatty acids, carbohydrates, and sterols. The amount and variety of available data is overwhelming, but has been underutilized due to a lack of tools with which to incorporate, manipulate, and correlate these data. This is the direction future work must take. Within the data analysis area, three needs remain unexplored or underdeveloped. The first is a need for tools that can embody multidimensional and very large data sets. As shown with the phospholipid data, none of the present tools can handle the complete phospholipid profiles that are easily obtained. Instead, data analysis must be done on subsets of the data, and only for a limited data base.

The second unmet need is for tools to prune and correlate data into sets describing separate qualities of an organism. For the phospholipids, the evidence is quite strong that the profiles contain information useful for taxonomic identification separate from that which correlates particular qualities of groups of organisms. In addition, the environmental factors experienced by a microorganism, or even the past history of a microorganism, can change the profiles enough so that pruning these data will be a necessity. Most likely, these types of behaviors occur within the data from all the chemical biomarkers. More importantly, these types of tools will allow us, through correlation of data, to uncover qualities of organisms previously unnoticed. Indeed, these tools could help solve many of the problems microbial taxonomists face in taxonomically placing new species or moving known species into a more taxonomically correct grouping.

Lastly, tools must be developed that incorporate polyphasic data. Phospholipid profiles are not the only data useful for microbial identification and correlation. The greatest strides can be made when all types of available information are used; including, for example, data available on all the chemical biomarkers, data obtained from carbohydrate utilization tests, and even data obtained from traditional microbiology tests. Limiting oneself to a single phase of data will never adequately describe the dynamics or subtleties of a microorganism.

Acknowledgments

This work was supported by a grant from the Center for Microbial Ecology, a National Science Foundation Science and Technology Center, Michigan State University. We thank Eric Hemenway (MSU) for development of the cluster analysis software and Edward Richter (Silliker Laboratories of Ohio) for cultures of the food-borne contaminants.

Literature Cited

(1) Ratledge, C.; Wilkinson, S. G. *Microbial Lipids, Volume 1*; Academic Press: San Diego, CA, 1988.
(2) Kates, M. In *Advances in Lipid Research, Volume 2*; Paoletti, R., ED.; Academic Press: New York, NY, 1964; Chapter 1.
(3) Lechevalier, M. P. *CRC Crit. Rev. Microbiol.* **1977**, *5*, 109-210.
(4) Heller, D. N.; Cotter, R. J.; Fenselau, C.; Uy, O. M. *Anal. Chem.* **1987**, *59*, 2806-2809.
(5) Pramanik, B. N.; Zechman, J. M.; Das, P. R.; Bartner, P. L. *Biomed. Environ. Mass Spectrom.* **1990**, *19*, 164-170.
(6) Heller, D. N.; Murphy, C. M.; Cotter, R. J.; Fenselau, C.; Uy, O. M. *Anal. Chem.* **1988**, *60*, 2787-2791.
(7) Cole, M. J.; Enke, C. G. *Anal. Chem.* **1991**, *63*, 1032-1038.
(8) Platt, J. A.; Uy, O. M.; Heller, D. N.; Cotter, R. J.; Fenselau, C. *Anal. Chem.* **1988**, *60*, 1415-1419.

(9) Cole, M. J.; Hemenway, E. C.; Enke, C. G. Paper presented at the 39th Annual Conference on Mass Spectrometry and Allied Topics, Nashville, TN, May 19-24, 1991.

(10) Bligh, E. G.; Dyer, W. J. *Can. J. Biochem. Physiol.* **1959**, *37*, 911-917.

(11) Zirrolli, J. A.; Clay, K. L.; Murphy, R. C. *LIPIDS* **1991**, *26*, 1112-1116.

(12) Huang, Z. H.; Gage, D. A.; Sweeley, C. C. *J. Am. Soc. Mass Spectrom.* **1992**, *3*, 71-78.

(13) Abel, K.; De Schmertzing, H.; Peterson, J. I. *J. Bacteriology* **1963**, *85*, 1039-1044.

(14) Brock, T. D.; Madigan, M. T. *Biology of Microorganisms, 5th ed.*; Prentice Hall: Englewood Cliffs, NJ, 1988; Chapter 19.

(15) Pfiffner, S. M.; Ringleberg, D. B.; Cole, M. J.; Tyndall, R. L.; White, D. C. manuscript in preparation.

(16) Jung, S.; Lowe, S. E.; Hollingsworth, R. I.; Zeikus, J. G. *J. Biol. Chem.* **1993**, *268(4)*, 2828-2835.

RECEIVED April 16, 1993

Chapter 5

Pyrolysis–Gas Chromatography–Mass Spectrometry

Detection of Biological Warfare Agents

A. Peter Snyder[1], Philip B. W. Smith[2], Jacek P. Dworzanski[3], and Henk L. C. Meuzelaar[3]

[1]U.S. Army Edgewood Research, Development, and Engineering Center, Aberdeen Proving Ground, MD 21010–5423
[2]Geo-Centers, Inc., Gunpowder Branch, Aberdeen Proving Ground, MD 21010–0068
[3]Center for Microanalysis and Reaction Chemistry, University of Utah, Salt Lake City, UT 84112

Curie-point wire and quartz tube pyrolysis (Py) devices coupled to short column (2 and 5 meter) gas chromatography/mass spectrometry (GC/MS) systems are investigated for their ability to extract microbiologically-useful information from unprocessed, underivatized whole microorganisms. The U.S. Department of the Army is sponsoring the development of a Chemical, Biological Mass Spectrometer (CBMS) system to autonomously operate in fielded situations. As such, data are presented that are relevant to the design of a fieldable device such as the CBMS system. Ubiquitous (e.g. nucleic acid bases) as well as more highly specific (e.g. certain lipids) biological constituents are addressed for their use in microorganism detection and characterization by Py-GC/MS. The need for rapid generation of useful information is presented as well as the desire for a simplified graphical representation suitable for direct interpretation as opposed to purely computerized "black box" pattern recognition methods.

Since the latter half of the 1980's, the U.S. Army Edgewood Research, Development and Engineering Center has undertaken the formidable task of addressing the issue of detection and identification of biological agents using analytical instrumentation. Further goals of this concept include: successful performance in outdoor scenarios; detection/identification within a few minutes; and a lightweight system. The biological agents include microorganisms (bacteria, fungi, rickettsia and viruses) as well as proteinaceous and non-proteinaceous toxins. This represents a significant additional direction from the traditional U.S. Army missions in the detection and identification of nerve and incapacitating chemical warfare (CW) agent compounds. CW agents are mainly of a low molecular weight and are discrete in nature, i.e., they are not chemically bound to elaborate organic or inorganic matrices. This latter fact allows for many straightforward chemical and analytical detection schemes since CW agents generally do not require fragmentation into individual molecular building blocks or functional groups and moieties. Thus, the task of detection and identification of CW agents can deal with the intact agent molecule as opposed to processed or altered molecules.

Biological compounds present a different situation and their detection requires a formidable leap with regard to analytical approaches and strategies. They usually exist as complex mixtures of low to very high molecular weight entities reaching many millions of atomic mass units. One such technology that appears promising in attaining the detection/identification goals of biological agents is mass spectrometry.

Because of the high molecular weight that characterizes microorganisms and given the mass limitations of conventional mass spectrometry, a sample processing module is necessary. This processing module must take the complex, high molecular weight biological agent and break it up into substances small enough to fall within the mass range of a small, field portable mass spectrometer.

A carefully weighed compromise between logistics (module size, consumables and power requirements), operating characteristics (speed of processing, sample size, sample collection efficiency, etc.) and scientific performance (sensitivity, specificity, overall information yield) eventually led to the choice of analytical pyrolysis techniques including so-called "oxidative" pyrolysis. Pyrolysis is usually defined as the rapid heating of a sample under an inert atmosphere with the objective of fragmenting the analyte into smaller, individual pieces. Oxidative pyrolysis, or the rapid heating of a sample under atmospheric or air conditions is the desirable alternative given the logistics requirements in that it eliminates the need for a bottled source of inert gas such as helium, nitrogen or argon.

This report will explore the interrelationships of the Curie-point wire and quartz tube pyrolysis (Py) concepts and the technique of mass spectrometry (MS) as well as its invaluable companion technology, gas chromatography (GC) as applied to the production of structural chemical information from microorganisms. Lipid components and nucleic acids are the principal biomaterials that are targeted in the bioanalytical characterizations.

Experimental Section

Curie-point Pyrolysis. A total of thirteen strains representing eight bacterial species were investigated. *Bacillus anthracis* (BA) (a virulent and low-virulent strain, B0463 and B0464, respectively), *Bacillus cereus* (BC) (B0037), *Bacillus thuringiensis* (BT) (B0158 and B0150), *Bacillus licheniformis* (BL) (B0017 and B0089), *Bacillus subtilis* (BS) (B0014 and B0095), *Staphylococcus aureus* (SA), and *Escherichia coli* (EC) (type 0127) were supplied by Tony P. Phillips (Chemical/Biological Defence Establishment, Porton Down, U.K.) and Leslie Shute (University of Bristol, Bristol, U.K.). *Legionella pneumophila* (LP) serogroup 1 was provided by Luc Berwald (Rijkinstituut voor de Volksgesondheid, Bilthoven, The Netherlands). Bacteria were grown in Lab M nutrient broth for 3 days at 37 °C. The cells were harvested by centrifugation, washed with sterile water, and then resuspended in water. Cells were killed by adding an equal volume of 6% H_2O_2 and incubating overnight. Samples were then centrifuged, washed, and freeze-dried. *L. pneumophila* (1 mg/mL suspension) were heat-killed at 120 °C and then lyophilized.

Suspensions of the organisms (1.7 mg/mL) were prepared by adding 0.15 mL of methanol to 0.5 mg of the lyophilized bacteria, sonicating to effect a uniform dispersion, and then adding 0.15 mL of deionized distilled water. A 3-*u*L sample, or approximately 5 *u*g of bacteria, was applied to the tip of a Curie-point wire, and the suspension was dried in a stream of warm (40 °C) air.

Py-GC/MS experiments were performed on a system including a Hewlett-Packard 5890 gas chromatograph (Palo Alto, CA) and a Finnigan-MAT ion trap mass spectrometer (San Jose, CA). The pyrolyzer consisted of a specially constructed low dead volume Curie-point pyrolysis GC inlet (*1-3*) and a 1-MHz, 1.5kW Curie-point power supply (Fischer 0310). The sample on the 610 °C Curie-point ferromagnetic wire was placed in the 280 °C pyrolysis head inlet and then heated to the Curie temperature within 1 sec. A 5 m x 0.32 mm I.D. capillary column coated with a 0.25-*u*m film of dimethylsilicone (J & W Scientific, SE-30) was operated with a He carrier gas at 175 cm/s linear velocity. The column was temperature-programmed from 100 to 300 °C at 40 °C/min and held isothermal for 5 min. The column was connected directly into the mass spectrometer vacuum through a 1 m transfer line maintained at 300 °C with the vacuum manifold at 180 °C. The ion trap operated without the normal Teflon spacer rings to allow better conductance of the high (8 mL/min) carrier gas flow.

The long-term reproducibility studies were performed on a Finnigan-MAT 700 ion trap detector system. Duplicate pyrolysis heads from that of the bacterial experiments were used with 610 °C Curie-point wires. A 5 m BP-1 column (Chrompack) with a 0.25-*u*m film thickness was used that had a helium linear velocity of approximately 175 cm/s. The GC inlet was at 8 psig, the transfer line was maintained at 280 °C, and the Model 8500 Perkin-Elmer GC system was operated from 100 to 300 °C at 30 °C/min. To accommodate the high gas flows, several holes were drilled into the spacers of the ion trap detector and the trap was held isothermal at 225 °C.

For both systems, spectra were accumulated over the mass range 100-620 m/z at a rate of 1 scan/sec for approximately 8 min. The Curie-point experiments were conducted under pyrolysis (nonoxidative) conditions with a He carrier gas.

Quartz Tube Pyrolysis. Two bacterial strains, *Bacillus anthracis* (B0463, a virulent strain) and *E. coli*, were investigated and were the same samples used in the Curie-point pyrolysis experiments.

The Py-GC/MS equipment consisted of a Pyroprobe Model 122 power supply with a platinum coil probe and a modified pyrolysis reactor/GC interface (Chemical Data Systems, Oxford, PA) on a Finnigan-MAT 700 ion trap (San Jose, CA). All pyrolyses were conducted in the pulsed mode, with a final set temperature of 1000 °C, and the total heating time was 20 sec. The bacterial samples (20-50 ug) were sandwiched between two quartz wool plugs in the quartz tube holder and the latter was inserted into the coil of the heating probe. The pyrolysis reactor/GC inlet system was maintained at 300 °C. The GC (5 m x 0.32 I.D.) capillary column coated with a 0.25 um chemically-bonded dimethylsilicone film (Alltech) was operated with helium carrier gas at 100 cm/sec linear velocity. The column was temperature programmed from 100 °C to 300 °C at 30 °C/min and then held at 300 °C. The column was connected directly into the mass spectrometer through a splitless manifold, and the 1 m portion of the capillary was maintained at 300 °C. Three 3.2 mm holes were drilled into the two Teflon spacer rings in the ion trap to compensate for the high carrier gas flow. Spectra were accumulated over a 100-650 amu range at a rate of 1 scan/sec for approximately 14 min.

Curie-point Pyrolysis with a Two Meter GC Column.   A 2 m total column length was used where one meter spanned the Curie-point wire to GC oven exit and the second meter comprised the transfer line to the ion trap mass spectrometer in a continuous (no-split) 2 m length of DB-5 column. The column ID was 0.18 mm with a 0.4 micron film thickness. Both sections of the column were held isothermal at 300 °C prior to pyrolysis. Also, to minimize condensation/precipitation on the inside of the glass reaction tube, a 358 °C ferromagnetic foil was carefully slipped inside the tube so as to cover the inside walls. The Curie-point wire was then inserted into the tube. The foil did not impede or touch the wire when it was positioned. Five micrograms of the virulent *B. anthracis* B0463 strain were placed on the 610 °C Curie-point wire from a liquid suspension. The GC column was operated at a 16 psi helium head pressure producing a column flow of 3 mL/min. The GC injector port that housed the GC column inlet was kept at 300 °C as was the entire pyrolysis reactor head.

The time axis (x-axis) of all mass chromatograms are presented in a minutes: seconds format.

Results and Discussion

The concept of pyrolysis-short column GC/MS using Curie-point and quartz tube pyrolysis devices was investigated for the ability to produce microbiologically-informative data from unprocessed microorganisms.

Shaw (4) has stated that of the myriad of components that can be found in microorganisms, the determination of their lipid composition is the closest to an ideal chemotaxonomic method. Lipids, along with nucleic acids, represent major components of all microorganisms. During pyrolysis, these classes of biochemical compounds from microorganisms tend to produce large, characteristic building blocks that can be readily analyzed by GC/MS. Although carbohydrate and protein moieties have also been reported to produce specific pyrolysis patterns permitting differentiation between selected microbiological species or even individual strains (5-10), lipid and nucleic acid patterns appear to have a much broader chemotaxonomic applicability. Consequently, the lipids and nucleic acids are the subject analytes of the present work.

Curie-point Pyrolysis.

Analytical Dimensions of Information.   Figure 1 presents selected analytical layers of information from a three-dimensional (retention time, m/z and intensity) dataset of the Py-GC/MS analysis of a low virulence B. anthracis strain, B0464.

Relatively sharp peaks characterize the first 3 min in the total ion chromatogram (TIC), while peaks in the region between 5 and 6 min are broader in nature (11). The latter region occurs at the high-boiling end of the chromatogram (260-300 °C), which is characteristic for diglyceride type lipid components.   Note that this area, or lipid TIC (vide infra) region, is well separated from earlier eluting materials.

The remainder of Figure 1 provides strong evidence for the lipid nature of the high-boiling components in the TIC and also portrays a diversity of masses and GC retention times suitable for organism differentiation purposes. The integrated, background-subtracted mass spectrum of the entire lipid TIC (lipid TIC mass spectrum) is shown below the lipid TIC region in Figure 1. Note in particular the 14 amu methylene repeat unit throughout most of the lipid TIC mass spectrum.

The reconstructed ion chromatogram (RIC) profiles of selected higher molecular weight ion species in Figure 1 are positioned directly above the TIC with respect to time. The RIC time axis coincides with the lipid TIC time axis. A RIC profile shows the intensity of a particular ion over the chromatographic elution profile.   Note that the higher the m/z value, the longer the retention time. Together, these and other higher molecular weight mass peaks produce distinctive TIC patterns.   The maximum intensity (or % of the TIC) of each RIC is plotted vs. m/z value and the contour is shown to the right of the RIC distribution in Figure 1.   This analytical dimension also provides a distinctive trend for pattern recognition purposes.

Finally, to the left of each RIC is the respective extracted ion mass spectrum. Note that for a given ion, a different pattern of masses is observed, both in the high mass region (m/z 450-600) and the middle mass region (m/z 250-350).   Combined, these analytical dimensions provide a useful format for visual differentiation between microorganisms.

Figure 2 provides a similar analysis for the E. coli organism.   Note in particular the abundance of lipids on the lower boiling side of the lipid TIC with respect to the B. anthracis organism in Figure 1.   Further differences can be observed in the RIC comparisons and RIC intensity vs. mass plots.   The equivalent extracted ion mass spectra for both organisms depict interesting

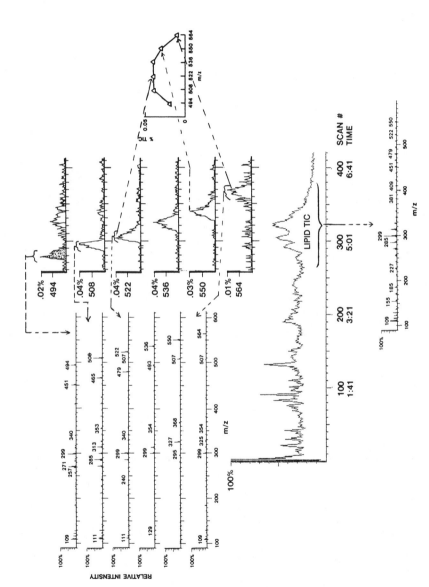

Figure 1. Curie-point Py-GC/MS analytical information in perspective for *B. anthracis* B0464. See text for details.

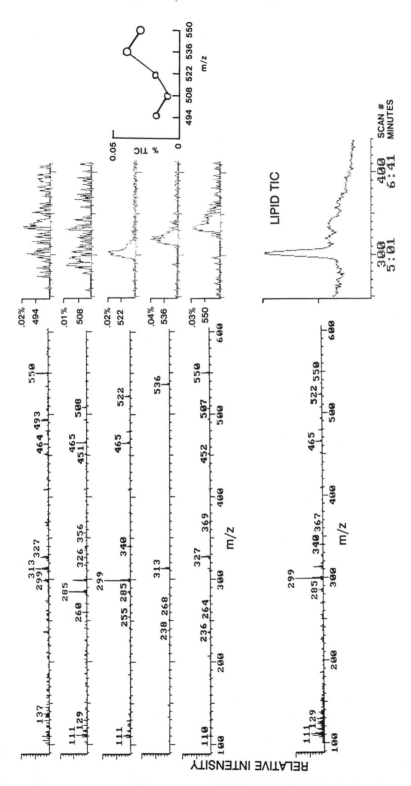

Figure 2. Curie-point Py-GC/MS analytical information in perspective for *E. coli*. See text for details.

differences which point to the fact that lipid signals common to both bacteria may well represent isomers as opposed to identical compounds.

Reproducibility. Short and long term reproducibility experiments of the Py-GC/MS profiles of *B. anthracis* B0463 were conducted (*11, 12*). The TICs in Figure 3a,b were acquired approximately five hours apart while Figure 3c was obtained five months later with a fresh bacterial suspension on a different Py-GC/MS system. The lipid TIC portion of all three experiments are very similar and the overall lipid TIC impression was detailed by very similar analytical layers of information (not shown) such as that in Figures 1 and 2.

Mislabeled Microorganism. Figure 4 shows the Py-GC/MS TICs of three organisms. The labels on the bottles containing the lyophilized powders that produced the Figure 4(a-c) spectra were *B. cereus* B0002, *B. licheniformis* B0017 and *B. cereus* B0037, respectively. It appeared that one of the bottles was mislabeled with respect to species. Since it is known that *B. cereus* is closely related to *B. anthracis* (*13-16*), the lipid TIC of Figure 4c is much more similar to that of the anthrax strain (Figures 1 and 3) than to that in Figure 4a. Therefore, by the lipid TIC envelope as well as the supporting mass spectral data (not shown), Figure 4a should represent a *B. licheniformis* strain as opposed to a *B. cereus* strain.

Nucleic Acid Information. The high molecular weight lipid information was investigated for their presence in a Py-GC/MS analysis of underivatized, unprocessed microorganisms. From the same experimental datasets, the earlier portion of the bacterial TICs was investigated for the presence of the low molecular weight nucleic acid nitrogen bases. Figure 5 represents RICs of the parent ion signals of adenine (m/z 135) and thymine (m/z 126) as well as of an adenine electron ionization fragment (m/z 108). The relative elution times are similar to that of a mixture of pure thymine and adenine (data not shown) where m/z 126 (thymine) elutes earlier than m/z 135 (adenine). These representative RICs are similar for all the bacteria as listed in the Experimental section. The remarkable observation of Figure 5 is that given the relatively low molecular weights of the three masses, these two nucleic acid bases dominate the bacterial pyrolyzate RICs. The most significant feature of the m/z 126 RIC is that the retention time is similar to that of pure thymine. Pure adenine Py-GC/MS features (m/z 135 and 108) produced similar RICs to that of m/z 135 and 108 in Figure 5. Many peaks arising from the myriad of pyrolyzate species from the bacteria were expected to be present in the RICs in Figure 5, however, this was not observed. Confirmation of the RIC identities were found in NBS mass spectral library first matches with the two compounds (Figures 6 and 7).

RIC Patterns of Uncertain Origin. The m/z 155 RIC for a variety of organisms is shown in Figure 8. For certain bacteria, a highly informative pattern is produced while for most other bacteria, very little is observed (*17*).

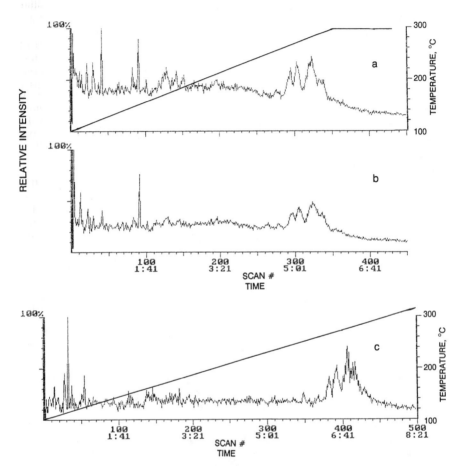

Figure 3. Py-GC/MS long term reproducibility experiments for *B. anthracis* B0463 (a-b) short term and (c) long term.

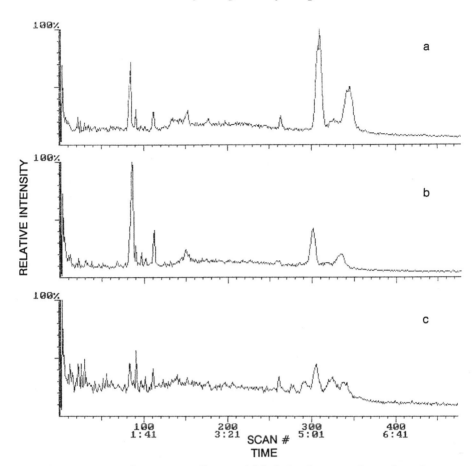

Figure 4. Experiments revealing a mislabeled microorganism. Samples were taken from bottles labeled as (a) *B. cereus* B0002, (b) *B. licheniformis* B0017 and (c) *B. cereus* B0037.

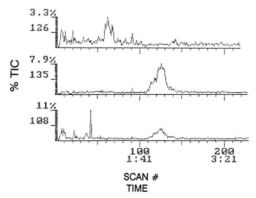

Figure 5. RICs of m/z 108, 126 and 135 from *B. anthracis* B0464.

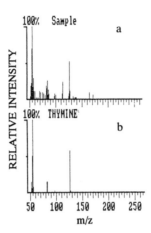

Figure 6. (a) Extracted m/z 126 mass spectrum (the average of mass scans 60-70 in Figure 5) from *B. anthracis* B0464 which produces a first match with the NBS library mass spectrum of pure thymine (b).

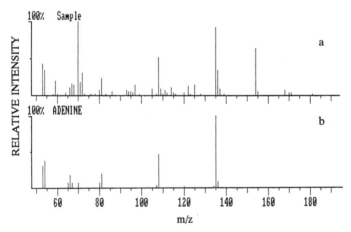

Figure 7. (a) Extracted m/z 135 mass spectrum (the average of mass scans 120-130 in Figure 5) from *B. anthracis* B0464 which produces a first match with the NBS library mass spectrum of pure adenine (b).

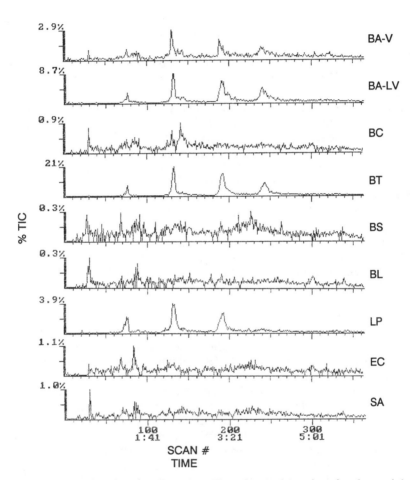

Figure 8. RICs of m/z 155. See Experimental section for bacterial abbreviations. V, virulent; LV, low virulent strain of *B. anthracis*.

The *B. anthracis* (BA) and *B. thuringiensis* (BT) bacteria produce similar patterns while *B. cereus* (BC) only depicts the first two peaks (70-80 and 130-150 sec), and *L. pneumophila* (LP) is essentially missing the last peak at 240 sec. The inclusion of this information presents an important point in bacterial differentiation. The extreme complexity of microorganisms includes well-documented biological substances. However, in analytical techniques such as Py-GC/MS, data and patterns can arise in which the biochemical origin of the features is unknown. Visually, the analysis is straightforward but the interpretation is unclear from a biochemical/microbiological point of view.

One Minute Generation of Bacterial Lipids. The generation of well-defined lipid TICs of various organisms with the Py-5m-GC/MS experiments prompted the question of reducing the time of appearance of the lipids. A reduction in the 4.5-6 min generation time was sought, thus, a 2m-300° C isothermal GC column was used. Figure 9 shows a *B. anthracis* chromatogram where the first 40 sec represents nonspecific eluate. The lipid TIC lies between 50 and 90 sec and is generally of a similar impression to that in Figures 1 and 3 performed under more "controlled" GC conditions. Therefore, it appears that under less stringent, less controlled GC conditions a measure of useful information from the high molecular weight, high-boiling lipids is possible.

An analysis of Figure 9 with respect to the lipid biochemical substances is presented in Figure 10 and follows a similar rationale as to that of the 5 m column Curie-point data (*11*). Note that in the extracted ion mass spectra, each mass spectrum is obtained from their respective RIC (Figure 10) at the indicated scan number, and an example is highlighted by the arrow. The "parent" or high mass peak increases every 14 amu from the m/z 494 to m/z 550 extracted mass spectrum. It is not practical to compare lipid TICs between the 5 m (Figures 1 and 3) and 2 m (Figure 9) experiments, because the latter has a severe biochemical background superimposed on the lipid signals. The RIC signals in Figure 10 show considerable overlap with each other, and compared to Figure 1 under a 5 m GC column, this is to be expected. A comparison of the 2m GC-derived extracted ion mass spectra and their respective equivalents with the 5 m GC condition portrays similarities and differences. The subtle distinctions in the 5 m GC column extracted ion mass spectra become less clear in the 2 m situation. Figure 11 portrays the results when a 10:90 w/w mixture of *B. anthracis* (B0463) and a dirt sample was pyrolyzed. The lipid TIC has fewer features than that in Figure 9, and the RICs are of lower intensity and more broadened (Figure 12). Nevertheless, Figure 12 shows that lipid mass spectral signatures can be observed. *E. coli* was also investigated by the 2 m GC column situation (Figure 13). Because *E. coli* has a more intense peak in its lipid TIC (Figure 2), this is also prominent in the 2 m GC situation (Figure 13) and is not significantly obscured by biochemical background signals. The lipid TIC mass spectrum is shown in Figure 13 and a high degree of similarity exists with the *E. coli* lipid TIC mass spectrum and its 5 m GC column equivalent (Figure 2).

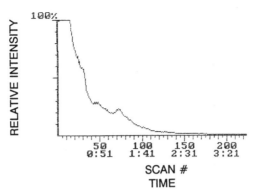

Figure 9.   One minute Py-GC/MS generation of the TIC of *B. anthracis* B0463.

Quartz Tube Pyrolysis. Quartz tube pyrolysis coupled to a GC/MS system was investigated in the processing of whole organisms (*18*). The high-boiling lipid TIC region was investigated in the analysis of *E. coli* and *B. anthracis*. The analyses followed the same investigative theme as that used in the parallel Curie-point Py-GC/MS experiments.

Quartz Tube Pyrolysis of Microorganisms. The lipid TIC obtained following the Py-GC/MS of a virulent strain of *B. anthracis* (B0463) is presented in Figure 14 and contains significant structural detail even in this first simple impression of the Py-GC/MS chromatogram.

A background-subtracted mass spectrum for the lipid TIC is presented in Figure 14. The essential features of the mass spectrum are similar to those in the non-virulent B0464 strain (Figure 1) except for the much higher intensities of the dehydrated diglyceride peaks in the quartz tube runs. A prominent feature of the spectrum is a series of ions differing by consecutive 14 amu increments. RICs for selected ions are presented in Figure 14 and as the parent ion increases in mass the retention time also increases. Several of the RICs, notably that for m/z 494 and 522, display a number of peaks; this can be attributed to the presence of isobaric compounds with different retention times. The mass spectral scans under the most intense portion of each RIC was where the average extracted ion mass spectrum was taken, and the mass spectra show that the highest molecular weight peak increases every 14 amu (left side of Figure 14) for each successive RIC (right side of Figure 14).

The results obtained for *E. coli* were treated in a similar fashion and are presented in Figure 15. The lipid portion of this organism is interesting in that m/z 494, 508 and 522 co-elute, hence, their extracted mass spectra overlap (confounded). The extracted mass spectrum between mass scans 390-395 reflects this phenomenon (Figure 15a). Note that m/z 522-523 dominates in intensity in the high mass region. This occurs because of their relatively high contribution to the total ion chromatogram (28%) relative to m/z 494-495 (1.0%) and m/z 508-509 (0.9%). The extracted mass spectra of m/z 536-537 (Figure 15b, mass scans 398-408) and m/z 550-551 (Figure 15c, mass scans 410-420) have no significant overlap with other major ions in their respective RICs.

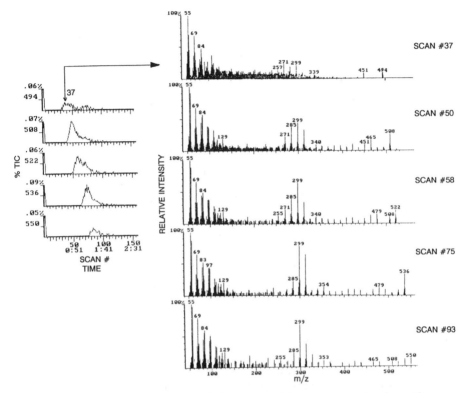

Figure 10. Selected RICs and their extracted ion mass spectra from the one minute Py-GC/MS analysis of *B. anthracis* B0463 (Figure 9). The reconstructed ion for each RIC is labeled on the vertical axis. The mass spectral scan number of each RIC is labeled and an example is shown for the m/z 494 RIC.

The extracted mass spectrum of the entire lipid portion of the TIC (Figure 15d, mass scans 370-450) reflects the dominance of m/z 522-523 and the relatively minor contribution of the other high mass ions.

A comparison between the Curie-point (Figure 2) and quartz tube (Figure 15) results of *E. coli* shows that similar mass spectral m/z distributions are present in each respective situation. Also, a high degree of similarity exists with respect to the chromatographic impressions of the lipid TICs and RICs of both sample processing techniques in the form of a visual pattern recognition.

Conclusions and Perspectives

Significant progress has been made in biological information generation and hardware development since the Mass Spectrometry Biological Detection

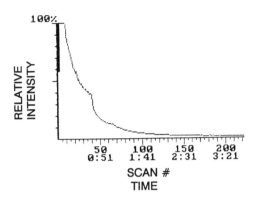

Figure 11. One minute TIC of Py-GC/MS analysis of a 10:90 mixture of
B. anthracis B0463 and a dirt sample.

Workshop was held at the U.S. Army Edgewood Research, Development and
Engineering Center in November 1985. Especially noteworthy events are the
rapid, straightforward elution of underivatized lipid material through
conventional GC columns into mass spectrometry devices after direct pyrolysis
of unprocessed microorganisms (*11, 12*). Computer algorithm software has
been developed to handle the immense amount of data that is necessarily
generated from pyrolysis investigations of complex compounds. More
importantly, a significant amount of biomarker compounds found in bacteria
were able to be identified by either pyrolytic or non-pyrolytic, analytical
microbiological methods (*8, 19-22*). These compounds are much more specific
than pre-1985 investigations in that they clearly are found in bacteria, in
general, or are specific for certain bacterial sets.

The current design of the autonomous, fieldable CBMS system employs
oxidative pyrolysis of samples with a quartz tube sample processor, and through
a few feet of heated fused silica transfer line, the GC eluate is analyzed by an
ion trap or Quadrupole Ion Storage Trap (QUISTOR) device. Another
significant element of this device is that the transfer line does not directly
empty eluate into the ion trap; rather, a polysilicone membrane resides between
the transfer line opening and the ion trap proper. The types of polar and non-
polar biomarker compounds that can diffuse through the membrane without too
much "band-broadening" are currently under investigation (*23*).

Pure microorganisms themselves provide a significant amount of
background biochemical noise when investigating for the presence of biological
marker compounds. For example, the Py-MS literature (*5, 24-31*) contains
many uninteresting, "useless" or "noisy" m/z signals, and the Py-GC/MS studies
(Figures 1-4, 9, 11, 13-15) show a significant offset in the chromatographic
baseline. When compared to the sheer mass and size of an organism, a
particular biochemical component pales in size and is similar to the "needle in
a haystack" problem.

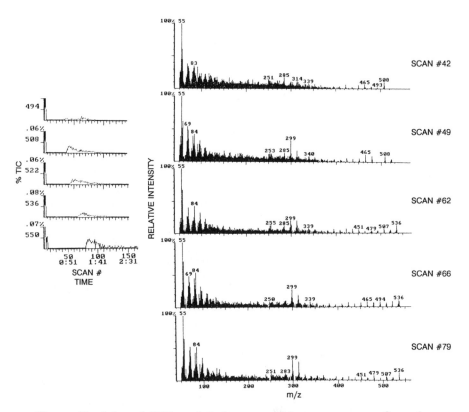

Figure 12.   Selected RICs and their extracted ion mass spectra from the one minute Py-GC/MS analysis of a mixture of *B. anthracis* B0463 and a dirt sample (Figure 11). See Figure 10 for details.

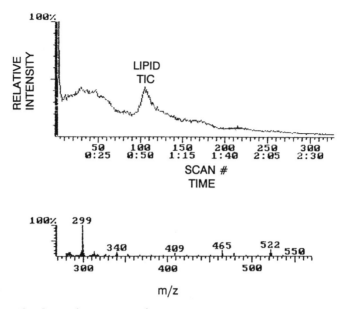

Figure 13.   One minute Py-GC/MS TIC and the lipid TIC mass spectrum of *E. coli*.

With respect to a stand-alone or portable/mobile Py-GC/mass spectrometer, information such as the presence of nucleic acid nitrogenous bases in a sample can allow for a decision-making algorithm to determine whether a substance of biological origin is present. It must be kept in mind that their presence can be due to plant and animal material as well as microorganisms, thus, this type of biomarker can serve as a generic piece of information to the analyst. However, the nucleic acid bases elute relatively early in a gas chromatographic analysis, and as such, their presence/absence can be determined prior to the later-eluting lipid material. If no bases are detected, then the analysis need not continue for lipid interrogation. Instead, another sample can be analyzed. If nucleic acid bases are reported, then the sample can continue to be analyzed until the GC lipid retention time/temperature conditions are reached.

The work of Voorhees et al. showed that in a complex mixture of bacteria and interferents (*32*), the biological signals could not adequately be resolved by statistical means. In simpler interferent matrices, Voorhees (*32*), Meuzelaar (*33*), P.B.W. Smith (unpublished data with quartz tube Py-GC/MS),

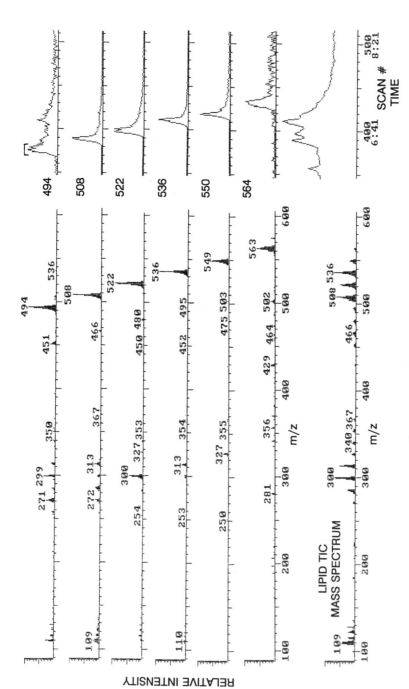

Figure 14. Quartz tube Py-GC/MS analytical information in perspective for *B. anthracis* B0463. Reconstructed ions are listed between each RIC-extracted mass spectrum pair. The bracket above the m/z 494 RIC indicates the region of its extracted mass spectrum.

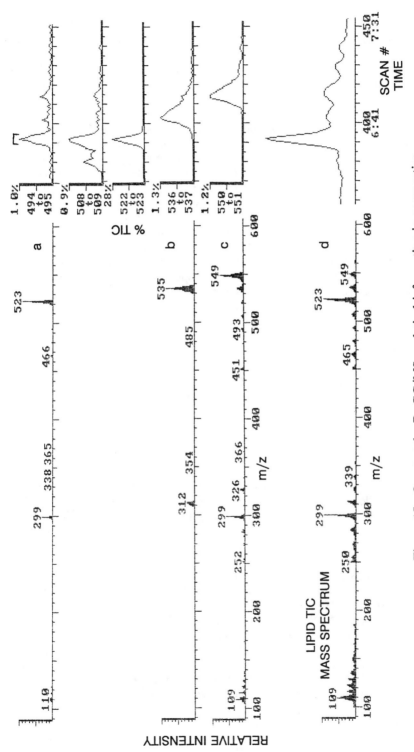

Figure 15. Quartz tube Py-GC/MS analytical information in perspective for *E. coli*. See Figure 14 for details.

and the present report showed that, by multivariate and visual methods, the sample (biological in nature) could be observed despite the interferential matrices. The issue of oxidative and non-oxidative pyrolysis with complex analytes such as microorganisms is an important point when minimization of logistics burdens is considered in an outdoor analysis scenario. A comparison of the Py-GC/MS lipid information derived from (non-oxidative) pyrolysis in the present report and references 11, 12 and 18 can be made with that of oxidative pyrolysis of microorganisms with both a Curie-point-MS (*34*) and quartz tube GC/MS (*35*) system. As a whole for the different organisms investigated, analogous GC patterns (TIC and RIC) information are observed for the Py-GC/MS studies and similar m/z distributions can be noted for all four permutations of oxidative and non-oxidative pyrolysis with Curie-point and quartz tube reactors. Thus, it seems possible to perform "air" pyrolysis of extremely complex analytes such as microorganisms and still have the benefit of a comparison of the results with the established pyrolysis literature.

In conclusion, more work, mainly in the fundamental discovery/research avenues, needs to be addressed with respect to pyrolyzate generation, pyrolyzate transfer to the mass spectrometry analyzer, and reduction of the complex data sets. Adequate instrumentation exists in order to improve on these principles.

Acknowledgments

The authors wish to thank Ms. Linda G. Jarvis for the preparation and editing of the manuscript and Richard R. Smardzewski for a critical reading of the manuscript.

Literature Cited

1.   Meuzelaar, H. L. C.; McClennen, W. H.; Yun, Y. *Proc. 35th ASMS Conference on Mass Spectrometry and Allied Topics*, **1987**, 1142-1143.
2.   McClennen, W. H.; Richards, J. M.; Meuzelaar, H. L. C. *Proc. 36th ASMS Conference on Mass Spectrometry and Allied Topics*, **1988**, 403-404.
3.   Richards, J. M.; McClennen, W. H.; Bunger, J. A.; Meuzelaar, H. L. C. *Proc. 36th ASMS Conference on Mass Spectrometry and Allied Topics*, **1988**, 547-548.
4.   Shaw, N. *Adv. Appl. Microbiol.* **1974**, *17*, 63-108.
5.   Medley, E. E.; Simmonds, P. G.; Manatt, S. L. *Biomed. Mass Spectrom.* **1975**, *2*, 261-265.
6.   Adkins, J. A.; Risby, T. H.; Scocca, J. J.; Yasbin, R. E.; Ezzell, J. W. *J. Anal. Appl. Pyrolysis* **1984**, *7*, 15-33.
7.   Smith, C. S.; Morgan, S. L.; Parks, C. D.; Fox, A.; Pritchard, D. G. *Anal. Chem.* **1987**, *59*, 1410-1413.
8.   Fox, A.; Gilbart, J.; Morgan, S. L. In *Analytical Microbiological Methods - Chromatography and Mass Spectrometry;* Fox, A.; Morgan, S. L.; Larsson,

L.; Odham, G., Eds.; Plenum Press: New York, NY, 1990, pp 1-17.

9.   Goodacre, R.; Berkeley, R. C. W.; Beringer, J. E. *J. Anal. Appl. Pyrolysis,* **1991,** *22,* 19-28.
10. Ford, T.; Sacco, E.; Black, J.; Kelley, T.; Goodacre, R.; Berkeley, R. C. W.; Mitchell, R. *Appl. Environ. Microbiol.* **1991,** *57,* 1595-1601.
11. Snyder, A. P; McClennen, W. H.; Dworzanski, J. P; Meuzelaar, H. L. C. *Anal. Chem.* **1990,** *62,* 2565-2573.
12. Snyder, A. P.; McClennen, W. H.; Meuzelaar, H. L. C. In *Analytical Microbiology Methods: Gas Chromatography and Mass Spectrometry;* Fox, A.; Morgan, S. L.; Larsson, L.; Odham, G., Eds.; Plenum: New York, NY, 1990, pp 201-217.
13. Logan, N. A.; Carman, J. A.; Melling, J.; Berkeley, C. W. *J. Med. Microbiol.* **1985,** *20,* 75-85.
14. Lawrence, D.; Heitefuss, S.; Seifert, H. S. H. *J. Clin. Microbiol.* **1991,** *29,* 1508-1512.
15. Turnbull, P. C. B.; Hutson, R. A.; Ward, M. J.; Jones, M. N.; Quinn, C. P.; Finnie, N.J.; Duggleby, C. J.; Kramer, J. M.; Melling, J. *J. Appl. Bacteriol.* **1992,** *72,* 21-28.
16. Logan, N. A.; Berkeley, R. C. W. In *The Aerobic Endospore-Forming Bacteria;* Goodfellow, M.; Jones, D.; Priest, F. G., Eds.; Academic Press: London, 1981, pp 105-140.
17. Snyder, A. P.; *Technical Report,* CRDEC-TR-342, Defense Technical Information Center, Cameron Station, Alexandria, VA   22304, April 1992.
18. Smith, P. B.; Snyder, A. P. *J. Anal. Appl. Pyrolysis* **1992,** *24,* 23-38.
19. Morgan, S. L.; Watt, B. E.; Ueda, K.; Fox, A. In *Analytical Microbiology Methods, Chromatography and Mass Spectrometry;* Fox, A.; Morgan, S. L.; Larsson, L.; Odham, G., Eds.; Plenum Press: New York, NY, 1990; pp 179-200.
20. Morgan, S. L.; Fox, A.; Gilbart, J. *J. Microbiological Methods* **1989,** *9,* 57-69.
21. Fox, A.; Udea, K.; Morgan, S. L. In *Analytical Microbiological Methods - Chromatography and Mass Spectrometry;* Fox, A.; Morgan, S. L.; Larsson, C.; Odham, G., Eds.; Plenum Press: New York, NY, 1990, pp 89-99.
22. Gilbart, J.; Fox, A.; Morgan, S. L. *Eur. J. Clin. Microbiol.* **1987,** *6,* 715-723.
23. Meuzelaar, H. L. C.; Kim, M. G.; Arnold, N. S.; Kalousek, P.; Snyder, A. P. *Proc. U.S. Army Research Office Workshop on Spectrometry and Spectroscopy for Biologicals,* **1991,** 38-50.
24. Meuzelaar, H. L. C.; Windig, W.; Harper, A. M.; Huff, S. M.; McClennen, W. H; Richards, J. M. *Science,* **1984,** *226,* 288-274.
25. Engman, H.; Mayfield, H. T.; Mar, T.; Bertsch, W. *J. Anal. Appl. Pyrolysis* **1984,** *6,* 137-156.
26. Irwin, W. J. *Analytical Pyrolysis;* Chromatographic Science: New York, NY, 1982; Vol. 22, Chapter 2, pp 45-89.
27. Voorhees, K. J. *Analytical Pyrolysis;* Butterworths: London, 1984.

28.   Freed, D. J.; Liebman, S. A. In *Pyrolysis and Gas Chromatography in Polymer Science;* Liebman, S. A.; Levy, E. J., Eds.; Chromatographic Science: New York, NY, 1985; Chapter 2, pp 15-51.
29.   Meuzelaar, H. L. C.; Haverkamp, J.; Hileman, F. D. *Pyrolysis Mass Spectrometry of Recent and Fossil Biomaterials: Compendium and Atlas;* Elsevier: Amsterdam, 1982.
30.   Boon, J. J.; Brandt-de Boer, B.; Eijkel, G. B.; Viegels, E.; Sijtsma, L.; Wouters, J. T. M. In *Mass Spectrometry in Biotechnological Process Analysis and Control;* Heinzle, E.; Reuss, M., Eds.; Plenum: New York, NY, 1987, pp 187-208.
31.   Simmonds, P. G. *Appl. Microbiol.* **1970,** *20,* 567-572.
32.   Voorhees, K. J.; Durfee, S. L. *Army Research Office Contractor Technical Report,* AD A176672, Research Triangle Park, NC, 1987.
33.   Meuzelaar, H. L. C.; Windig, W. *Final Contractor Report to Army Research Office,* Contract DAAG29-84-K-0009, Research Triangle Park, NC, 1987.
34.   DeLuca, S.J.; Voorhees, K.J. *J. Anal. Appl. Pyrolysis* **1993,** *24,* 211-225.
35.   Smith, P.B.; Snyder, A.P. *J. Anal. Appl. Pyrolysis* **1993,** *24,* 199-210.

RECEIVED May 14, 1993

Chapter 6

Detection of Grain Infection with Specific Toxicogenous Fungal Species

Gas Chromatography–Mass Spectrometric Analysis of Volatile Metabolites

V. N. Emokhonov[1], I. N. Groznov[1], O. A. Monastirskii[2], and A. Yu. Permogorov[1]

[1]Institute of Energy Problems of Chemical Physics, Russian Academy of Sciences, Moscow, Russia
[2]North Caucasus Research Institute of Phytopathology, Krasnodar, Russia

Microscopic fungi in their metabolism produce specific volatile organic substances, which can be detected with gas analysis methods. Using GC-MS analysis of headspace samples taken over grain infected with specific toxinogenous fungi strains we have characterized some of these compounds. For Fusarium and Piricularia strains certain volatile organic mixtures were detected that are specie specific. These volatile complexes may be considered as indicators of the pathogen infecting grain and may present the possibility of a sensitive, fast one stage method to monitor toxinogenic infections of specific fungi in stored grain.

Harvested cereals are always at risk for infection by any of a large number of microscopic fungi and bacteria, different species of Aspergillum, Penicillim and Fusarium, for example. Grain may be spoiled by such phytopathogenic microorganisms while it is being improperly transported or stored. Such infections diminish the quality of the grain and result in weight loss, however the greater danger comes from infection with fungi that produce toxins that are harmful to warm-blooded animals, including man. Aspergillum species produce aflatoxins; Fusarium species produce zearalenon, desoxyvalenol and T-2 toxin. Rapid diagnosis of infection of grain by these fungi in the early stages is very important in order to identify shipments that should be isolated or destroyed, to recognise and correct inadequate storage conditions, and to identify shipments that should be processed and used quickly.

Methods used to monitor grain for infections of toxinogenic fungi include organoleptic analysis, microscopic examination of scrapings and washings, culturing of colonies of microorganisms out of grain samples, and extraction of toxins for bioassay or physico-chemical analysis. However, detection of infections in their early stages by these methods is difficult, because these methods are so

0097–6156/94/0541–0085$06.00/0

slow and labor intensive. To test a single grain sample for infection by toxinogenic fungi classically requires a separate añlysis for each possible pathogenic species, and a separate analysis for each toxin potentially present. Grain stored in large bins is usually not infected uniformly, but rather in one or more limited sites. Consequently, numerous solid samples must be taken, from different parts of the bin, and each subjected to multiple analyses. It has been recognised in this laboratory and others that the analysis of volatile metabolites from the enclosed atmosphere above stored grain can provide an analysis integrated over the entire volume of grain.

The analysis of specific volatile organic substances (odor volatiles) produced by metabolic processes of toxinogenic fungi may provide the basis of a rapid method that will allow detection of grain infections in a single process. In this approach samples of the atmosphere (headspace) over grain stored in a closed bin are collected for examination by standard physico-chemical methods. Some of the volatile fungal metabolites associated with moldy grain have been identified by Kaminski et al (1,2), Stawicki et al (3), and Abramson et al (4,5). Tuma, Sinha and co-workers (6,7) showed that the amount of these odor volatiles detected in headspace samples taken from storage bins correlates with the extent of microfloral infection in the grain. Unfortunately, the volatile metabolites identified in these studies, namely 3-methyl-1-butanol, 3-octanone and 1-octen-3-ol, appear to be non-specific with respect to the fungus species present. Thus these compounds can be used only as indicators of general infection, and not as indicators of specific species of fungi or of the presence of toxinogenic populations.

Consequently, the objective of this study was the characterization of volatile compounds that are the by-products of specific toxinogenic fungal species and that are easily analyzed with standard methods. These could then be used as indicators of grain infections by particular species.

METHODOLOGY

Samples of damp rice were sterilized in 1 liter flasks and infected separately with live mycelium of the following species of toxinogenic fungi: Aspergillus flavus, Pyricularia oryzae, Fusarium macroceros, Fusarium sporotrichiella, and two strains of Fusarium graminearum. Experimental and uninfected control flasks were incubated at 20-22° C. Head space samples were taken for analysis at 7, 15, 30 and 50 days after infection. Similar experiments were carried out with 100 liter stainless steel storage bins containing 25 kg of rice and 25 kg of wheat, infected with Fusarium graminearum.

Headspace samples were collected by passing approximately 0.5 liter of air through a stainless steel tube filled with TENAX TA sorbent. The probes were thermally desorbed into a cryogenic trap coupled to a capillary column in an interfaced gas chromatograph ion trap mass spectrometer. A standard volume of saturated naphthalene vapor was added to each sorbent tube as an internal standard for quantitation based on the areas of chromatographic peaks. Volatile organic compounds were tentatively identified (8,9) by matching their mass spectra against those in the library of electron impact mass spectra of 60000 compounds compiled by the National Institute for Science and Technology (USA).

RESULTS

Only volatile compounds present in samples of air over infected grain and absent in air over control samples were considered as possible indicators of grain deterioration. In this way any odors that might have penetrated from the ambiant atmosphere could be excluded. A number of byproducts of grain infection were detected. Table 1 presents tentative structures and estimated concentrations of 12 of these odor volatiles detected in air samples taken 15 days after infection of rice, as well as information on their presence in a control sample.

In addition to the compounds observed in all of the deteriorated grain headspace samples, several odor volatiles are detected only in samples from grain infected with species of Fusarium. These compounds may serve to indicate infection of grain by specific pathogenic fungi. Unfortunately satisfactory library matches could not be obtained for spectra of these compounds, however most of them can be classified as sesquiterpenes or their derivatives. Similar compounds, in particular caryophyllene, were reported by Tuma et al (7) in air samples taken from experimental storage bins containing wheat infected with a large number of microflora species, including Fusarium. However, in that study these volatiles were not attributed to microfloral infection.

Figure 1 presents part of the total ion chromatogram of the headspace sample from rice infected with Fusarium macroceros, taken 15 days after infection. More than 100 peaks were detected in the complete capillary chromatogram, most of which were also present in the control analysis. Chromatographic peaks of Fusarium-specific volatiles are marked in Figure 1 by numbers that correspond to those in the first column of Table 1. Mass spectra are also presented of two of these chemical markers (Figure 2 and 3).

The Fusarium-specific volatiles were also present in head space samples taken 7, 30 and 50 days after infection. Their levels of concentration stayed roughly constant past 30 days, and were seen to be decreasing at 50 days, when fungal vitality had diminished overall. All of the Fusarium species produce qualitatively similar mixtures, however the relative concentrations of individual components differ quantitatively. Because of the large experimental error, estimated as +- 15%, proportions of specific components could not be correlated chemotaxonomically.

Experiments done in the large scale storage bins revealed the same specific mixtures of volatile compounds for both rice and wheat infected with Fusarium.

A unique compound was also found in the air over rice infected with Pyricularia. It is unidentified and designated number 7 in Table 1.

Mass spectrometry used in combination with gas chromatography provided a method to identify many of the common and specific volatile compounds by library searching. Mass spectrometry enhanced chromatographic detection by providing deconvolution of overlapping GC peaks. In subsequent development of this screening method, selected ion monitoring may permit the use of more rapid chromatographic separations without loss of specificity.

Our studies indicate that infections with Fusarium and Pyricularia may be distinguished from each other and from Aspergillus by analysis of the volatile metabolites they form. The possibility now exists for a fast and sensitive method

Table 1. Odor Volatiles in Air Samples from Rice 15 Days
after Infection with Different Species of Fungi

Concentration, ng/l[#]

Compound	1	2	3	4	5	6	7
1H-Imidazole,2-Methyl	41	464	146	337	527	809	200
2-Butenal, 2-Ethenyl	3	2	5	40	413	9	6
Octynone,Dimethyl	-	-	-	7	26	-	-
1H-Imidazole,2-Ethyl-4-Methyl	-	-	3	7	74	-	-
Heptadienal,2,4-Dimethyl	-	1	-	-	4	-	-
Unidentified 1	-	-	-	42	36	26	29
Unidentified 2	-	-	-	173	327	14	15
Unidentified 3	-	-	-	39	50	35	24
Unidentified 4	-	-	-	483	2753	16	11
Unidentified 5	-	-	-	164	257	24	26
Unidentified 6	-	-	-	67	444	160	35
							1
							2
Unidentified 7	-	-	4	-	-	-	-

1-Control uninfected grain
2-Aspergillus flavus
3-Pyricularia oryzae
4-Fusarium macroceros, strain 240
5-Fusarium sporotrichiella, strain 417
6-Fusarium graminearum, strain 16/2
7-Fusarium graminearum, strain 4/1

estimated uncertainty ± 15%

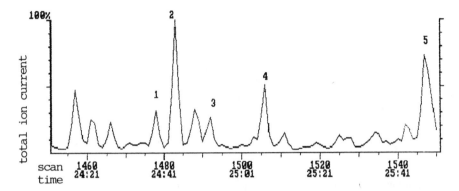

Figure 1.   A portion of the total ion chromatogram of an air sample taken after 7 days from the vessel containing rice grain, infected with Fusarium graminearum. Specific odor volatiles peaks are marked by numbers that correspond to those in Table 1.

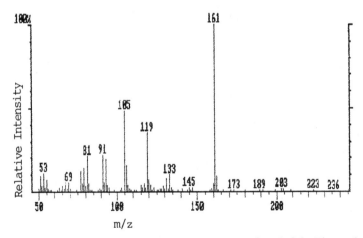

Figure 2.   Electron impact mass spectrum of peak 3 in Figure 1.

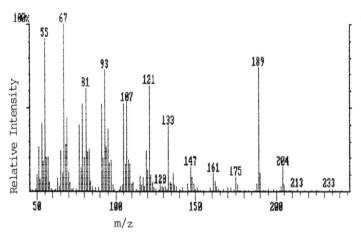

Figure 3.   Electron impact mass spectrum of peak 5 in Figure 1.

to monitor toxinogenic infections of specific fungal spcies in stored grain, and is the subject of ongoing research in this laboratory.

Literature Cited

1. Kaminski,E., Stawicki,S., Wasowicz,E. and Przybilski,R. *Ann. Technol. Agric.* **1973** (Paris) 22, 261.
2.Kaminski,E., Stawicki,S. and Wasowicz,E. *Appl. Microbiol.* **1974,** 27,1001.
3. Stawicki,S., Kaminski,E., Newiarowitz,A., Troyan,M. and Wasowicz,E. *Ann.Technol. Agric.* **1973** (Paris) 22, 309.
4. Abramson,D., Sinha,R.N. and Mills,J.T. *Cereal Chem.* **1980,** 57,346.
5. Abramson,D., Sinha,R.N. and Mills,J.T. *Cereal Chem.* **1983,** 60,350.
6. Sinha, R.N., Tuma,D., Abramson,D. and Muir,W.E. *Mycopathologia* **1988,** 101, 53.
7.Tuma,D., Sinha,R.N., Muir,W.E. and Abramson,D. *Int. J. Food Microbiol.* **1989,** 8,103.
8. *Gas Chromatogr./Mass Spectrom. Appl. Microbiol.* Eds. Larsson, L., Mardh, P.A., Odham, G. Plenum: New York, N.Y. **1984,** 207.
9. Edwards, R.A., Dainty, R.H., Hibbard, C.M. *J. Bacteriol.* **1987,** 62, 403.

RECEIVED September 27, 1993

Chapter 7

Direct Detection of Volatile Metabolites Produced by Microorganisms

Membrane Inlet Mass Spectrometry

F. R. Lauritsen[1] and D. Lloyd[2]

[1]Department of Biochemistry, Odense University, Campusvej 55,
DK–5230 Odense M, Denmark
[2]Microbiology Group (PABIO), University of Wales College of Cardiff,
P.O. Box 915, Cardiff CF13TL, Wales, United Kingdom

Volatile metabolites produced by microorganisms can be measured directly in growth media by membrane inlet mass spectrometry (MIMS). A polymer membrane, which allows volatile compounds to pass, is used as the only separation between the mass spectrometer and the liquid to be analyzed. The practical aspects of membrane inlet design are reviewed and recent innovations, e.g. the use of microporous membranes with solvent chemical ionization are discussed. In addition, applications of MIMS to the direct monitoring and identification of volatiles in microbial cultures are described.

Membrane inlet mass spectrometry has become a popular technique for the direct measurement of gases and volatile organic compounds in aqueous solution. The technique has been used especially for on-line measurements of dissolved gases in bioreactors, kinetic studies of chemical and biological reactions, identification of volatiles produced by microorganisms and environmental monitoring. All these applications have been reviewed (*1-6*). Recently a new method was developed (*7,8*), where the vaporized solvent of the samples is used as reaction gas for chemical ionization. With this technique detection of small polar compounds at low ppb concentration levels was possible. In addition, recent work indicates that even non volatile compounds such as glucose might be measurable with the technique in the near future (*9*). The possible application field of MIMS is therefore rapidly growing, and the identification of microorganisms based on MIMS measurements of metabolites is one such application where MIMS might be advantageous, as compounds can be measured directly in the growth medium without any pre-treatment.

The requirements for specificity, and in some special cases for sensitivity in measurement of microbially-produced volatiles has emphasized the need for the introduction of hyphenated techniques into biotechnology. The exacting specification desired of such methods is highlighted by consideration of the complexity of

0097–6156/94/0541–0091$06.00/0

fermentation mixtures, *e.g.* yeast is known to produce at least 111 products including, 22 secondary alcohols, 12 aldehydes, 12 α-keto acids, 28 esters, 22 free acids and 13 other carbonyl and phenolic compounds (*10*). Examples of the use of GC/MS are too numerous to be exhaustively cited here, and we mention just two types of application. (i) Aroma and flavor compounds important in the food industry are routinely measured by this method. Many microbial (especially fungal) flavors are of commercial interest (*11-13*). (ii) Differences between closely related strains can provide useful taxonomic criteria. Concentration of headspace vapors by chemical or cryogenic trapping (*14*) or solvent extraction from aqueous media is a normal prerequisite for GC/MS analysis.

Spoilage of food products associated with off-odours has also been investigated by headspace gas analysis (*e.g.* vacuum-packed beef spoilage by *Clostridium sp.*, (*15,16*). A comparison of volatiles produced by 31 strains of pseudomonads isolated from meat stored in air at chill temperatures with those produced by reference strains of *Ps. fragi* and *Ps. fluorescens* biotype I (*17*) amply illustrates the great discriminatory potential of GC/MS for identification purposes as well as the use of this technology in food science. A fast GC/MS/MS method for headspace analysis has been described by van Tilborg (*18*).

In its present stage of development MIMS does not offer the same specificity and sensitivity as GC/MS, but recent results indicate that this might happen in the near future. MIMS is advantageous by comparison with GC/MS in the handling of samples, since measurements can be carried out directly on the samples without any pre-treatment.

Membrane Inlets and Their Construction

Membrane inlets exist in many different designs, each design suited for a particular application or measuring problem. However, all designs have to deal with the same basic transport problems, i.e. diffusion of sample through an unstirred liquid layer (Nernst layer) in front of the membrane, transport through the membrane and transport of the vaporized sample from the inside surface of the membrane to the ionization region of the mass spectrometer. In the design of a membrane inlet one must compromise between an ideal solution to the transport problems and practical considerations regarding the particular application. The membrane inlets in common use today can be divided into four different categories: 1. Membrane probes mounted directly in a reactor (Figure 1a) (*2,19-20*); 2. Stirred sample cells (Figure 1b) (*21-25*) 3. Flow cells (Figure 1c) (*26-29*); 4. He-purge inlets (Figure 1d) (*30*); The characteristics and applications of the different inlets will be discussed in the section on membrane inlet design and application.

Transport through Polymer Membranes. Only compounds that can absorb into the membrane, pass through it and evaporate from the inside surface of the membrane into the vacuum can be detected by MIMS. Therefore the choice of membrane is decisive for the variety of compounds that can be measured. Most of the inlets use a non-porous membrane. The flow through such membranes follows Fick's diffusion and the steady state flow can be described as

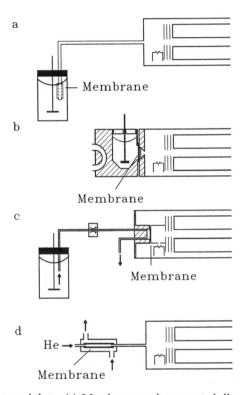

Figure 1. Membrane inlets. (a) Membrane probe mounted directly in a reactor. (b) Sample cell mounted close to the ion source of the mass spectrometer. (c) Flow cell with a sample stream passing by a membrane mounted in or at the ion source. (d) He purge inlet where the inside of a silicone capillary is continuously purged with helium, while aqueous sample flows continuously over the exterior of the membrane.

$$I_{ss} - A \cdot S \cdot D \cdot \frac{p}{1} \qquad (1)$$

where I_{ss} is the steady state flow rate through the membrane (mol/s), A is the membrane area (cm^2), S is the solubility constant (mol/torr cm^3), D the diffusion constant (cm^2/s), p the vapor pressure of the analyte on the sample side of the membrane (torr) and 1 is the membrane thickness (cm). Both the solubility constant and the diffusion constant have a strong dependance on molecular properties of the sample and the possibility of enrichment of certain compounds as compared to the solvent is evident. This principle is used in most membrane inlet systems, where hydrophobic membranes are used for measurements in aqueous solution and hydrophilic membranes for measurements in organic solutions. The common hydrophobic membranes are Teflon, polyethylene and polypropylene for the selective measurement of gases (1), and silicone polymers for the detection of gases and hydrophobic, volatile organic compounds (31); hydrophilic polyethylene terephthalate is used for the selective detection of water (32). The selectivity of the membrane is both an advantage and a disadvantage. The selectivity of the membrane simplifies the mass spectra obtained a great deal, and very simple instrumentation is adequate for most monitoring applications, but at the same time the diversity of compounds that can be measured is limited. In aqueous solution, where silicone rubber membranes are used, low detection limits (ppb) are only obtainable for hydrophobic compounds and the detection limits increase with the polarity of the compounds (5).

Porous Membranes. The flow through a porous membrane can be regarded as a combination of laminar/turbulent flow and simple diffusion, depending on the pore size of the membrane. The porous Teflon and polypropylene membranes that have been used in connection with MIMS (7-8, 25) have relatively large pore sizes (0.02 μm) and the expected molecular cut-off value is several hundred thousand Daltons. A large part of the transport of compounds through these membranes must therefore be expected to be by laminar or turbulent flow, and only minor enrichment of samples by comparison with the solvent can be expected. This hypothesis was supported in recent experiments with porous membranes (8), where similar detection limits for a variety of organic compounds having widely different hydrophobicities were found. The results further indicate that evaporation from the membrane surface rather than transport through the membrane is the limiting factor for the variety of compounds that can be measured with MIMS when porous membranes are used.

Membrane Response Time. Response time is a very important factor for an on-line monitoring technique. For nonporous membranes the 10-90% response time can be expressed as (5)

$$t_{10-90} - 0.237 \cdot \frac{1^2}{D} \qquad (2)$$

This equation shows that the response time is inversely proportional to the diffusion constant, which depends on both molecular properties, membrane material and temperature. The important molecular parameter is molecular size, since small molecules diffuse much faster than large molecules. In many cases, however, interactions between sample molecules and membrane material can give unexpectedly fast response times. For example this is often the case for esters and ketones when hydrophobic silicone membranes are used (*33*). With an elevation of temperature the diffusion constant increases and the response time becomes shorter. On average the diffusion constant increases 50% when the temperature is raised from 25 to 50 °C (*34*). In general the response time will be of the order of a few minutes for organic compounds which can be measured with silicone membranes. Differences in response times can be used to identify unknown compounds in mixtures. Peaks which appear in the mass spectrum with the same transient might belong to the same compound, whereas peaks which appear with different transients must belong to different compounds (*24*).

With porous membranes very fast response times ($\approx$ 100 ms) have been reported for gases and some small organic compounds (*25*) and in general response times for larger compounds have been reported to be of the order of tens of seconds (*7,8*).

Transport through Vacuum Tubes. In cases where evacuated tubes (high vacuum) are used to connect the membrane inlet with the ion source of the mass spectrometer slow transients are to be expected as a result of adsorption/desorption processes on vacuum surfaces (*28*). For many years this problem prevented measurements of less polar compounds (boiling point > 200 °C). However, with the development of direct insertion membrane probes, where the membrane is mounted inside (*26,27*) or in the immediate vicinity (*28,29*) of the ion source, the problem has been greatly reduced. Temperature has a very pronounced effect on the vacuum transient as demonstrated by Bohátka et al. (*3*) in an experiment where the response time of propanol through a 7 m long vacuum tube was shown to be reduced from 25 hours at 34 °C to less than 10 min at 90 °C.

Transport through the Unstirred Liquid Layer. Whenever a liquid is flowing past a stationary surface it must have zero velocity in the immediate vicinity of the surface and an unstirred liquid layer results (the Nernst layer, *35*). This causes a problem if the surface is a device which consumes certain compounds. The partial pressure of these compounds at the surface will depend on the rate at which the compounds are consumed by the device, the rate at which the compounds can diffuse through the unstirred liquid layer, and on the rate at which the compounds are produced or consumed in the unstirred layer (*36*). In most applications of MIMS the problem of the unstirred liquid layer can be eliminated simply by fast stirring of the solution or by fast flushing of the membrane with the sample liquid. Only in applications where the flow around the membrane is very slow, e.g. in physiological studies where membrane probes are inserted into the blood stream of an animal, could the unstirred liquid layer create a problem.

Membrane Inlet Design and Application. Capillary probe membrane inlets inserted directly into reactors (Figure 1a) have been used for gas measurements in many applications. These inlets which can be used with all kinds of membranes are connected to the mass spectrometer through a long vacuum tube (20 cm - 7 m). Stirring in the reactor is normally sufficient to ensure a minimal Nernst layer in front of the membrane, and transport through the membrane of gases and small volatile organic compounds is relatively fast. The applicability of these membrane inlets is limited by condensation effects in the vacuum tube, which create slow transients or prevent organic compounds from reaching the ion source of the mass spectrometer. In physiology, membrane probes with a diameter less than a millimeter (37,38) have been inserted directly into the blood vessels of animals and humans. In this way on-line and in-vivo studies of blood gas levels have been possible. In biotechnology, membrane probes have been inserted into fermentors for on-line monitoring of dissolved gases and small volatile organic compounds (3-4,20) and the measured values have been used for feed-back control (39-41). A major advantage in these applications has been the sterilization of the membrane probe prior to the measurements. These inlets have also been used for photosynthetic and ecological studies (42-44).

The configuration of choice for most kinetic studies consists of a membrane covered port through one wall of a sample cell mounted in the vicinity of the ion source of the mass spectrometer (Figure 1b). Typically these reaction vessels have a volume between 1 and 10 ml and must be equipped with a mechanical stirrer. Thermostating of the inlet is also necessary. Since the sample cell can be mounted close to the ion source of the mass spectrometer most volatile organic compounds can be measured in the absence of disturbing condensation effects in connection tubes. The membranes used for these inlets are sheet membranes of Teflon, polyethylene or polypropylene for selective gas measurements, and silicone rubber for the measurement of organic compounds. Measuring cells have been designed in many different variations, among which are cells for photosynthetic studies (21), cells open for gas exchange for steady state kinetic studies (22,45), cells combined with stopped flow mixing for the measurement of fast transients (23,46-47). Sample cells are also useful for the identification of metabolites produced by microorganisms. The growth medium (or a filtrate thereof) is simply transferred to the measuring cell, and after a few minutes mass spectra and/or tandem mass spectra can be recorded (24). In electrochemistry measuring cells have been developed with the aim of measuring very fast response times (< 1 s). Here microporous membranes and differential pumping of the ion source are used to achieve a fast flow through the system (25).

The largest diversity of organic compounds that can be measured with a membrane inlet is obtained with flow cells (Figure 1c). In these designs a continuous stream of sample passes a membrane in a small cell mounted as close to the ion source as possible. The optimal design has the membrane mounted inside (26,27) or at (28,29) the ion source. This reduces problems with condensation effects to virtually zero and limitation is reduced to transport through the membrane and in some cases to evaporation from the membrane surface (8). Stirring over the membrane surface is ensured by the sample flow through the measuring cell. Flow cells have been very successfully used for the measurements of environmentally important pollutants such as chlorocarbons, benzenes and aldehydes (48-51). They have also been used for the study of chemical reactions such as chlorination of amines (52). In these applications a peristaltic pump is used to pump the reaction liquid through the flow cell. After passage through the flow cell the reaction liquid is returned to the reactor. This setup can also be used for the identification of metabolites produced by microorganisms, when a filtrate of the growth medium is continuously circulated through the flow cell (*53*). A major advantage of flow cells is that samples can be chemically modified before measurements in order to create volatile compounds (*54*). The combination of flow injection analysis sampling and a flow cell has been especially useful for the on-line monitoring of fermentation products (*55*)

A new type of membrane inlet was recently introduced (*30*), where the interior of a silicone rubber hollow fiber is continuously purged with helium, while aqueous sample flows continuously over the exterior membrane surface. The helium purge gas removes sample molecules from the inside of the hollow fiber and carries the sample molecules to the ion source. With this inlet, ppt detection limits were obtained for some hydrophobic compounds.

Instrumentation. It is very easy to connect membrane inlets to a mass spectrometer and they have been used with quadrupole and triple quadrupole mass spectrometers, ion traps and magnetic sector instruments. The complexity of the instrument depends on the application. For most monitoring purposes where the composition of the sample is known a small quadrupole mass spectrometer with electron impact ionization is sufficient. The analysis of samples for the confirmation or identification of compounds requires the use of chemical ionization and/or tandem mass spectrometry.

Recent and Future Techniques. In the past, almost all measurements of organic compounds in aqueous solution have been carried out with silicone membranes. This is due to the large enrichment of sample achieved for hydrophobic compounds which dissolve very well in silicone rubber membranes. Polar compounds on the other hand do not dissolve easily in silicone rubber membranes and only a minor enrichment of sample is obtained. Porous membranes show a similar problem since little or no enrichment of sample is achieved during the transport through these membranes. The problem with volatile polar compounds was recently solved (*7,8*) by the use of solvent chemical ionization. In this technique a microporous membrane is used to create a sufficiently high flow of both solvent and analyte into the ion source of the mass spectrometer for chemical ionization conditions to be established with the evaporated solvent acting as reagent gas. A much higher total flow of compounds into the mass spectrometer compensates for the lack of enrichment and detection limits for polar, volatile organic compounds in aqueous solution was improved by 1-2 orders of magnitude by comparison with silicone rubber membranes. The technique of solvent chemical ionization can be used both with aqueous and organic solvents, although the high water pressure obtained with aqueous solutions reduce the filament lifetime to just a few hours and the use of glow discharge is necessary (*8*). At present the limitation of the technique is not in the transport of compounds through the membrane but in the evaporation of sample molecules from the evacuated surface of the membrane. However, techniques using different kinds of stimulated evaporation from the membrane surface are under development at Odense University, Denmark and Purdue University, IN. Experiments have actually been carried out (56) where glucose was measured directly from an aqueous solution by the use of hydrophilic and porous membranes, but further work still has to be done to produce stable and quantitative signals. However, these recent developments give much hope that MIMS in the nearest future will allow the direct measurement of larger and polar compounds such as penicillin and glucose directly from microbiological media.

Applications of MIMS to the Direct Detection and Monitoring of Volatiles in Microbial Cultures.

Yeasts. Outlet gas from a yeast fermentation was first monitored on-line by Pungor *et al.* (57). Using silicone rubber membrane, and using N_2 as an internal calibration standard, ethanol (as well as O_2 and CO_2) were measured continuously. For ethanol, m/z 46 (EI-MS) values gave good correlation between broth concentrations and head space gas provided that instrument calibration at m/z 28 was corrected for a small contribution from CO^+. Carbon balances of 96% and good correlation between CO_2 and ethanol production rates confirmed the validity of the measurements. Transient behavior could not be analyzed because of poor instrument response times. Parallel studies on yeast fermentations were also reported at this time (58,59). Heinzle *et al.* (60) performed continuous and simultaneous on-line analyses of these fermentation products plus acetaldehyde in gas (capillary inlet) and liquid phases (125 μm silicone membrane) in a 7-liter continuous culture of yeast. Transients occurring when process conditions were altered (gas flow rates or dilution rates) could be faithfully recorded, and oscillations in dissolved gases and in ethanol concentrations were observed. The ethanol signal required daily calibration by reference to GC, and difficulties were experienced in the analysis of acetaldehyde due to its low concentration and the similarities in fragmentation patterns of the two compounds.   Scale-up to production scale fermentors (100 m³) used for the production of yeast biomass from molasses in an aerated fed-batch system has been described by Cox (61). Reasonable agreement between ethanol measurements at m/z 31 and off-line chemical determinations was obtained, and the response time ($t_{90\%}$ = 3 min for a 60 cm long 0.6 cm i.d. stainless steel probe at ambient temperatures) was quite adequate for this process.

Continuous monitoring of ethanol in a stirred system open for gas flow provides a useful laboratory method for whole cell studies of yeast physiology (62). A short (10 cm) quartz probe (3.5 mm o.d), sealed at one end, and with a 0.25 mm hole 5 mm from the sealed end covered with a stretched silicone rubber sleeve (o.d. 0.19 mm, i.d. 0.15 mm) in a working volume of 1 ml gave a response time ($t_{50\%}$) of 0.7 min (m/z 31 at 30°C). This configuration has enabled a study of the Pasteur effect (the inhibition of glucose utilization by O_2) in a variety of glycolytic mutants of *S.cerevisiae* (63). More than 100 mM ethanol accumulates during the initial glucose-repressed phase of aerobic growth with medium initially containing 55 mM glucose. In many strains other products (acetaldehyde, citrate, glycerol, succinate and malate measured by HPLC) are by comparison with ethanol only minor ones. Acetic acid can accumulate under some conditions in some yeast cultures (*e.g.* in those of *Brettanomyces* spp.) and, at the low pH values characteristically employed for yeast fermentations, also penetrates the silicone membrane to provide a contribution at m/z 31. Ethanol oxidation under aerobic conditions is rapid in most yeast strains examined, and this must be taken into account when calculating rates of production. Where rapid response times for ethanol are required, continuous flow through an external loop over a heated flat silicone rubber (Esco, thickness 100 μm) membrane gives excellent results (*e.g.* $t_{90\%}$ = 20 s at 80°C; (*64*)).

The yeast *Brettanomyces anomalous* is highly resistant to food preservatives and sometimes contaminates fermented beverages. It produces strong spicy, smoke-like, clove-like, woody or phenolic odours leading to spoilage of wine, beer and cider (*65*). When cultures were maintained unshaken in the dark for 21 days in the presence of either coumaric acid, or ferulic acid, they accumulated volatile products (*66*). Identification of 4-ethylphenol and 4-ethyl-2-methoxyphenol was by GC/MS after a lengthy extraction and concentration procedure. The power of membrane inlet tandem mass spectrometry for rapid characterization of such compounds directly in the growth medium was recently demonstrated by Lauritsen et al (*24*). They used a simple membrane inlet triple quadrupole mass spectrometer, which they had designed and built for on-line fermentation control in cooperation with VG Quadrupoles (Cheshire, UK) . The instrument, which was described in detail elsewhere (*24*), is similar in principles to the early instrument of Yost and Enke (*67*). It was equipped with a sample cell similar to the one illustrated in Figure 1b. *Brettanomyces anomalous* was grown for 21 days in the presence of either coumaric acid, ferulic acid or syringic acid as precursors for metabolic products. Filtrates of the growth media were then transferred to the measuring cell and after 10 minutes EI and CI mass spectra and tandem mass spectra were recorded. Incubation of the organisms with ferulic acid yielded abundant ions in the CI-MS spectra at m/z 89 and 153, incubation with coumaric acid yielded abundant ions at m/z 89 and 123, whereas incubation with syringic acid just yielded m/z 89. Figure 2 a,b and c shows collision induced dissociation product spectra of the protonated ions at m/z 153, 123 and 89 respectively. The product ion spectra of m/z 89 were similar in all samples. The products causing the CI-MS ions at m/z 153 from the sample grown in the presence of ferulic acid and m/z 123 from the sample grown in the presence of coumaric acid were identified as 4-ethyl-2-methoxyphenol and 4-ethylphenol respectively, confirming the earlier GC/MS findings by Heresztyn (*66*). The CI-MS ion at m/z 89 found in all samples, except controls grown without addition of precursors, was identified as ethyl acetate not previously identified as a product from these precursors. The concentration of ethyl acetate was estimated as 350 μM in all samples.

Protozoa.   Ethanol is also a major product in some protozoa that   live in environments where oxygen is present at low levels or is totally absent. Thus MIMS has been used to study the effects of aeration on ethanol formation in a mutant strain of the cattle parasite *Tritrichomonas foetus* (*68*).   For the human gut parasite, *Giardia lamblia* (*69-71*) continuous measurements of ethanol at m/z 31 (EI-MS) were made simultaneously with fluorimetric monitoring of cytosolic NAD(P)H in washed cell suspensions.   By this means it was shown that ethanol production rates are highly sensitive to the intracellular redox state, as metabolic switching between alternative pathways terminating in alanine or acetate can occur.

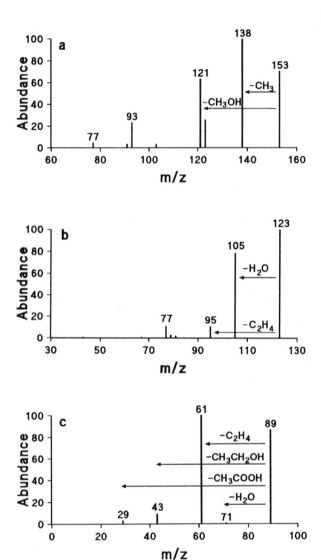

Figure 2. Collision-induced dissociation product spectra of growth media from *Brettanomyces anomalous*. (a) m/z 153 from sample grown with ferulic acid; (b) m/z 123 from sample grown with coumaric acid; (c) m/z 89 from sample grown with syringic acid. Adapted from ref. 24.

Using the membrane inlet triple quadrupole mass spectrometer described in the section on yeasts Lloyd et al. *(72)* studied the accumulation of volatiles in the growth media of trichomonad cultures. The complex medium used to cultivate these parasites contains not only yeast extract and peptone, but also 10% (v/v) horse serum. The EI-MS spectrum (Figure 3a) of those constituents of the medium which are volatile and penetrate 25 µm-thick silicone membrane showed peaks at m/z 58, 128, 72, 105, 106, 77, 76 and 64. That these peaks correspond to utilizable nutrients is indicated by their low abundance or absence in the spectra recorded after 1 day growth (Figure 3b) and 4 days growth (Figure 3c) of the human parasite *Trichomonas vaginalis.* CI-MS of the 4 days old culture showed peaks at m/z 118, 117 and 95. Collision induced dissociation product spectra of the $[M+1]^+$ parent ion m/z 118 gave fragments at m/z 91 and 65 and comparison with standard spectra gave unequivocal assignment of indole. The product ion spectra of the protonated ion at m/z 95 showed peaks at 79 and 61 and standard spectra indicated dimethyl disulphide. The concentration of indole and dimethyl disulphide was estimated as 20 µM and 4 µM respectively and together they accounted for all major peaks in the EI-MS spectrum of the 4 days old sample. Culture supernatants from two different clinical isolates of *Trichomonas vaginalis* strain 1910 and UCH-1 showed almost identical spectra, whereas the parasite *Tritrichomonas foetus* isolated from cattle only showed production of indole.

Bacteria.   Since the classical study of Weaver and Abrams, who in 1979 demonstrated the potential usefulness of silicon rubber as a selectively permeable barrier to low mol. wt volatiles (*e.g.* formaldehyde, methanol, acetaldehyde, ethanol, ethylene glycol, dimethyl sulphoxide and pyridene), these properties have been exploited in investigations of bacterial fermentations which often result in the accumulation of highly complex mixtures of organic compounds. Thus Doerner *et al.* *(73)* were able to perform on-line monitoring of ethanol, acetone, n-butanol, acetic acid and butyric acid in cultures of *Clostridium acetobutylicum.* This study emphasized some of the pitfalls of a simplistic approach using EI mass spectrometry, where overlapping cracking patterns of different components make resolution difficult, and where high mixed solvent concentrations may give cross-interacting membrane transport coefficients. Heinzle *et al.* *(74)* reported on-line measurement of volatiles in cultures of *Bacillus subtilis, Streptomyces* spp. and *Corynebacterium dipheriae* using factor analysis for the interpretation of mass spectra during the progress of the fermentation. Acetone and butanediol have been measured in *B.subtilis* cultures *(75)*, and this study stressed the importance of good mixing for oxidized product formation. The materials used for probe construction influence measured response times. Disappearance of propanol, a carbon source used for erythromycin production, from cultures of *Streptomyces erythreus* was measured by use of stainless steel probes connected to the MS analyzer by steam-jacketed lines in pilot-scale reactors *(3,76-77)*; the response time ($t_{90\%}$ of 8 min at 90°C in a 7 m pipe-run was quite appropriate for fermentations that span 5 days.

  Klebsiella oxytoca culture supernatants were analyzed through a dimethyl vinyl silicone capillary tube located within 1 mm of the electron beam of the ion source of a Finnigan MAT TSQ 4500 triple quadrupole mass spectrometer *(26)*. Collision induced dissociation product spectra from m/z 91 and 89 ions in CI mass

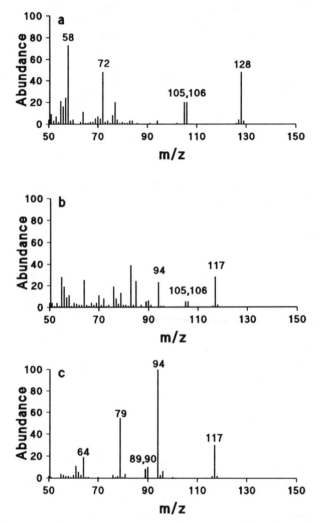

Figure 3. Electron impact mass spectra of growth medium from *Trichomonas vaginalis* strain UCH-1. (a) Medium before inocculation; (b) Medium after 1 day growth; (c) Medium after 4 days growth.

spectra (methane) confirmed the identities of 2,3-butanediol and 3-hydroxy-2-butanone (acetoin) in the fermentation broth; ethanol was also present. High sensitivity and fast response times were achieved by close proximity of the membrane to the heated ionizing region. Fall times (defined as the time taken after flow-through at 0.5 ml/min for the signal to fall from 90% to 10% of the maximum) were measured for 2-pentanone at high source temperatures (*e.g.* at 190°C which gave a value of 31.7s).   On-line monitoring of *K.oxytoca* and *B.polymyxa* fermentations was reported by Hayward *et al.* (*55*).   A continuous flow probe constructed with a flat dimethylvinylsilicone membrane, and with controlled electrical heating, enabled flow injection analysis of fermentation broth samples interspersed with standards.   Tandem mass spectrometry confirmed that measurements of acetic acid, acetoin, 2,3-butanediol and ethanol (m/z 61, 89, 91 and 47 after isobutane chemical ionization) were possible without cross-interference. Collision activation spectra under multiple ion conditions (with Ar target gas) were used to confirm product identities and to identify unassigned peaks (*e.g.* those from acetone and formic acid) in the CI spectra. Further work on optimization of the performance of the continuous flow probe (especially with respect to temperature and flow rates) makes an important contribution to bioreactor monitoring techniques (*5,29*).   Reaction network modelling using ongoing on-line rate calculations can provide information for on-line control of fermentation to improve cell mass and butanediol production (*78*).   The use of a microporous polypropylene membrane interface instead of silicone rubber in the continuous flow probe (*7,8*) for the direct identification of volatile organic compounds in either organic or aqueous solutions promises an extension to the range of fermentation products that can be monitored on-line.

Conclusion

Since the invention of membrane inlet mass spectrometry almost thirty years ago, the technique has evolved from a simple technique for analysis of dissolved gases into a more general technique for the analysis of volatile organic compounds in both aqueous and organic solution. The technical development has been followed by a steady increase in the number of applications, with environmental monitoring and identification of metabolites produced by microorganisms as recent examples. The number of publications where the technique has been used for direct measurements of metabolites produced by microorganisms is rapidly growing, a direct consequence of the ease by which samples are handled. At present the literature has no examples of the use of metabolites measured by MIMS for the identification of microorganisms, but the examples of the opposite application, identification of metabolites produced by known microorganisms indicate promising possibilities for such purposes. The most recent results, which indicates that the measurement of larger biomolecules such as penicillin and glucose might be possible in the near future, could make MIMS a unique method for the rapid identification of microorganisms based on measurements of produced metabolites.

Literature Cited

1. Woldring, S.; *J. Assoc. Advan. Med. Instr.* **1970** *4*, 43.
2. Lloyd, D.; Scott, R.I.; *J. Microbiol. Methods* **1983**, *1*, 313.
3. Lloyd, D.; Bohátka, S.; Szilágyi, J. *Biosensors* **1985** *1*, 179.
4. Degn, H.; Cox, R.P.; Lloyd, D. *Methods Biochem. Anal.* **1985** *31*, 165.
5. Kotiaho, T.; Lauritsen, F.R.; Choudhury, T.K.; Cooks, R.G.; Tsao, G.T. *Anal. Chem.* **1991** *63*, 895A.
6. Degn, H. *J. Microbiol. Methods* **1992** *15*, 185.
7. Lauritsen, F.R.; Kotiaho, T.; Choudhury, T.K.; Cooks, R.G. *Anal. Chem.* **1992** *64*, 1205.
8. Lauritsen, F.R.; Choudhury, T.K.; Dejarme, L.E.; Cooks, R.G. *Anal. Chim. Acta* **1992** *266*, 1.
9. Kotiaho, T.; Lauritsen F.R.; Choudhury, T.K.; Dejarme, L.E.; Srinivasan, N.; Cooks, R.G., Purdue University, Indiana, Unpublished results.
10. Suomalainen, H. *Suomen Kemist.* **1968** *41*, 239.
11. Schindler, J. *Ind. Eng. Chem. Prod. Res. Dev.* **1982** *21*, 537.
12. Berger, R.G.; Neuhäuser, K.; Drawert, F. *Biotechnol. Bioeng.* **1987** *30*, 987.
13. Gross, B.; Gallois, A.; Spinnler, H.-E.; Langlois, D. *J. Biotech.* **1989** *10*, 303.
14. Kallio, H. *J. Chromat. Sci.* **1991** *29*, 438.
15. Edwards, R.A.; Dainty, R.H. *J. sci. Food Agric.* **1987** *38*, 57.
16. Dainty, R.H.; Edwards, R.A.; Hibbard, C.M. *J. Sci. Food Agric.* **1989** *49*, 473.
17. Edwards, R.A.; Dainty, R.H.; Hibbard, C.M. *J. Bacteriol.* **1987** *62*, 403.
18. van Tilborg, M.W.W.M. In *Mass Spectrometry in Biotechnological Process Analysis and Control*; Heinzle, E.; Reuss, M., Ed.; Plenum Press, New York, 1987, 179-186.
19. Woldring, S.; Owens, G.; Woolford, D.C. *Science* **1966** *153*, 885.
20. Reuss, M.; Piehl, H.; Wagner, F. *European J. Appl. Microbiol.* **1975** *1*, 323.
21. Hoch, G.; Kok, B. *Arch. Biochem. Biophys.* **1963** *101*, 160.
22. Lundsgaard, J.S.; Petersen, L.C.; Degn, H. In *Measurement of Oxygen*; Editors, Degn, H.; Balslev, I; Brook, R.G., Ed.; Elsevier, Amsterdam, **1976**, 168.
23. Degn, H.; Kristensen, B. *J. Biochem. Biophys. Methods* **1986** *12*, 305.
24. Lauritsen, F.R.; Nielsen, L.T.; Lloyd, D.; Bohátka, S.; Degn, H. *Biol. Mass Spectrom.* **1991** *20*, 253.
25. Willsau, J.; Heitbaum, J. *J. Electr. Acta* **1986** *31*, 943.
26. Bier, M.E.; Cooks, R.G. *Anal. Chem.* **1987** *59*, 597.
27. Sturaro, A.; Doretti, L.; Parvoli, G.; Cecchinato, F.; Frison, G.; Traldi, P. *Biomed. Mass Environ. Mass Spectrom.* **1989** *18*, 707.
28. Lauritsen, F.R. *Int. J. Mass Spectrom. Ion Process.* **1990** *95*, 259.
29. Bier, M.E.; Kotiaho, T.; Cooks, R.G. *Anal. Chim. Acta* **1990** *231*, 175.
30. Slivon, L.E.; Bauer, M.R.; Ho, J.S.; Budde, W.L. *Anal. Chem.* **1991** *63*, 1335.

31. Westover, L.B.; Tou, J.C.; Mark, J.H. *Anal. Chem.* **1974** *46*, 568.
32. Bohátka, S.; Degn, H. *Rapid Commun. Mass Spectrom.* **1991** *5*, 433.
33. Lauritsen, F.R.; Bohátka, S.; Degn, H. *Rapid Commun. Mass Spectrom.* **1990** *4*, 401.
34. Lapack, M.A.; Tou, J.C.; Enke, C.G. *Anal. Chem.* **1990** *62*, 1265.
35. Nernst, W. *Z. Phys. Chem.* **1904** *47*, 52.
36. Lundsgaard, J.; Grønlund, J.; Degn, H. *Biotech. Bioeng.* **1978** *20*, 809.
37. Brantigan, J.; Gott, V.; Martz, M. *J. Apl. Physiol.* **1972** *32*, 276.
38. Lundsgaard, J.; Jensen, B.; Grønlund, J. *J. Appl. Physiol.* **1980** *48*, 376.
39. Joergensen, L.; Degn, H. *Biotechnol. Lett.* **1987** *9*, 71.
40. Whitmore, T.N.; Lloyd, D.; Jones, G.; Williams, T.N. *Appl. Microb. Biotechnol.* **1987** *26*, 383.
41. Paget, T.A.; Lloyd, D. *Molec. Biochem. Parasitol.* **1990** *41*, 65.
42. Miller, M.; Pedersen, J.Z.; Cox, R.P. *Biochim. Biophys. Acta* **1988** *943*, 501.
43. Lloyd, D.; Davies, K.J.P.; Boddy, L. *FEMS Microb. Ecol.* **1986** *38*, 11.
44. Lloyd, D.; Ellis, J.E.; Hillman, K.; Williams, A.G. *J. Appl. Bact. Symp. Suppl.* **1992** *73*, 1558.
45. Degn, H.; Lauritsen, F.R. *J. Phys. Chem.* **1989** *93*, 2781.
46. Gaunt, D.M.; Degn, H.; Lloyd, D. *Yeast* **1988** *4*, 249.
47. Carlsen, H.N.; Degn, H.; Lloyd, D. *J. Gen. Microbiol.* **1991** *137*, 2879.
48. Harland, B.J.; Nicholson, P.J.D.; Gillings, E. *Water Res.* **1987** *21*, 107.
49. Lister, A.K.; Wood, K.V.; Cooks, R.G.; Noon, K.R.; *Biomed. Environ. Mass Spectrom.* **1989** *18*, 1063.
50. LaPack, M.A.; Tou, J.C.; Enke, C.G. *Anal. Chem.* **1991** *63*, 1631.
51. Choudhury, T.K.; Kotiaho, T.; Cooks, R.G. *Talanta* **1992** *39*, 573.
52. Kotiaho, T.; Hayward, M.; Cooks, R.G. *Anal. Chem.* **1991** *63*, 1794.
53. Lauritsen, F.R.; Kotiaho, T; Lloyd, D., Odense University, Denmark, unpublished results.
54. Weaver, J.C.; Abrahms, J.H. *Rev. Sci. Instrum.* **1979** *50*, 478.
55. Hayward, M.J.; Kotiaho, T.; Lister, A.K.; Cooks, R.G.; Austin, G.D.; Narayan, R.; Tsao, G.T. *Anal. Chem.* **1990** *62*, 1798.
56. Kotiaho, T.; Lauritsen, F.R.; Choudhury, T.K.; Dejarme, L.E.; Srinivasan, N.; Cooks, R.G., Purdue University, Indiana, unpublished results.
57. Pungor, E.; Perley, C.R.; Cooney, C.L.; Weaver, J.C. *Biotechnol. Lett.* **1980** *2*, 409.
58. Weaver, J.C.; Perley, C.R.; Reames, F.M.; Cooney, C.L. *Biotech. Lett.* **1980** *2*, 133.
59. Weaver, J.C.; Perley, C.R.; Cooney, C.L. *Enzyme Enging.* **1980** *5*, 85.
60. Heinzle, E., Furukawa, K., Dunn, I. J. & Bourne, J. R. *Biotechnology* **1983** *1*, 181.
61. Cox, R.P. In *Mass Spectrometry in Biotechnological Process Analysis and Control*; Heinzle, E. & Reuss, M., Ed.; Plenum Press, New York, 1987, pp 63-74.
62. Lloyd, D.; James, C. J. *FEMS Microbiol. Lett.* **1987** *42*, 27.
63. Lloyd, D.; James, C. J.; Maitra, P. K. *Yeast* **1992** *8*, 291.

64. Lloyd, D., Cardiff, unpublished results.
65. Peynaud, E.; Domercq, S. *Arch. Mikrobiol.* **1956** *24*, 266.
66. Heresztyn, T. *Arch. Microbiol.* **1986** *146*, 96.
67. Yost, R.A.; Enke, C.G. *Anal. Chem.* **1979** *51*, 1251A
68. Lloyd, D.; James, C. J.; Lloyd, A. L.; Yarlett, N; Yarlett, N.C. *J. Gen. Microbiol.* **1987** *133*, 1181.
69   Lloyd, D.; Yarlett, N.; Hillman, K.; Paget, T.A.; Chapman, A.; Ellis,J.E.; Williams, A.G. In *Biochemistry and Molecular Biology of "Anaerobic" Protozoa;* D. Lloyd; Coombs G.H.; Paget T.A.; Ed.; Harwood, Chur, 1989, pp. 1-21.
70. Paget, T.A.; Raynor, M.H.; Shipp, D.W.E.; Lloyd, D. *Molec. Biochem. Parasitol.* **1990** *42*, 63.
71. Paget, T.A.; Kelly, M.L.; Jarroll, E.L.; Lindmark, D.G.; Lloyd, D. *Molec. Biochem.Parasitol.* **1993** *57*, 65.
72. Lloyd, D., Lauritsen, F.R.; Degn, H. *J. Gen. Microbiol.* **1991** *137*, 1743.
73. Doerner, P.; Lehmann, J.; Piehl, H.; Megnet, R. *Biotechnol. Lett.* **1982** *4*, 557.
74. Heinzle, E., Kramer, H. & Dunn, I. *J. Biotechnol. Bioenging.* **1985** *27*, 238.
75. Griot, M.; Heinzle, E.; Dunn, I.J.; Bourne, J.R. In *Mass Spectrometry in Biotechnological Process Analysis and Control*; Heinzle, E. & Reuss, M., Ed.; Plenum Press, New York, 1987, pp 75-90.
76. Bohátka, S. In *Gas Enzymology*; Degn, H.; Cox, R.P.; Toftlund, H., Ed.; Reidel, Dordrecht, 1985, pp 1-16.
77. Bohátka, S.; Szilágyi, J.; Langer, G. In *Mass Spectrometry in Biotechnological Process Analysis and Control*; Heinzle, E. & Reuss, M., Ed.; Plenum Press, New York, 1987, pp 115-124.
78. Austin, G.D., Tsao, G.J., Hayward, M.J., Kotiaho, T., Lister, A. & Cooks, R.G. *Process Control Qual.* **1991** *1*, 117.

RECEIVED May 14, 1993

Chapter 8

Identification and Detection of Carbohydrate Markers for Bacteria

Derivatization and Gas Chromatography–Mass Spectrometry

Alvin Fox and Gavin E. Black

Department of Microbiology and Immunology, School of Medicine, University of South Carolina, Columbia, SC 29208

Bacterial polysaccharides contain a diverse collection of unique sugar monomers which can serve as markers to identify specific bacterial species or genera. We commonly perform analyses employing hydrolysis followed by the alditol acetate derivatization procedure. This procedure involves destruction of the anomeric center, with sodium borohydride or borodeuteride, followed by acylation with acetic anhydride. In this instance the sugars are present in sufficient abundance that when analyzed by GC-MS (after electron impact ionization) in the total ion mode these compounds can be readily identified. In complex matrices such as airborne dust (or animal tissues and body fluids) markers for bacteria are present at sufficiently low concentration/absolute abundance that selective ion monitoring and extensive clean-up is vital for detection. Substantial improvements have been made including simplification and automation of certain parts of the alditol acetate procedure. Possible developments in alternative MS strategies for identification and detection of bacterial sugars are also discussed.

Analytical microbiology is a relatively new discipline that has arisen on the interface between analytical chemistry and microbiology. In the 1960's and 70's microbial chemistry established the composition of many major structural components of bacteria. These structures include peptidoglycans, lipopolysaccharides and teichoic acids. The unique monomers present within these compounds as well as their natural variability has provided a framework for taxonomic characterization of bacterial species. Instrumental developments in the 1980's, such as the commercial availability of fused silica capillary GC columns, as well as relatively inexpensive and simple-to-maintain bench-top mass spectrometers, has allowed clarification and expansion of our knowledge on the chemical composition of microbes. These technological advances have provided a means by which unique microbial compounds can be used for trace detection of

0097–6156/94/0541–0107$07.25/0

bacteria in complex matrices having environmental, biotechnology and medical implications.

Bacterial Carbohydrates

Bacterial species (prokaryotes) are sufficiently diverse, both genetically and biochemically that they are grouped into two kingdoms (archaebacteria and eubacteria). All established bacterial pathogens are eubacteria. Other cellular forms of life, including plants and animals, are grouped in a single kingdom (eukaryotes). Mammals, including man, are just one small group within this kingdom. Chemical components that are unique to or prominent in certain groups of microorganisms (e.g species, genera, families), are termed chemical markers. Characteristic structures found in bacteria include unique sugars, fatty acids and amino acid monomers. The development of simple, selective and sensitive analytical chemical methods for these markers offers an alternative and supplement to traditional biochemical and microbiological methods (1).

   Some markers (such as muramic acid, D-alanine or β hydroxy myristic acid) are present in high concentrations in many eubacteria, but are uncommon in higher life forms such as plants or animals (2-5). Other markers (such as aminodideoxyhexoses common among the family, *Legionellaceae*) (6,7) or D-arabinitol (a metabolite of *Candida*) are less widely distributed among different microbial species or genera (8). Gas chromatography-mass spectrometry (GC-MS) can be used to differentiate whole bacterial cells or detect microbial contaminants in complex samples as diverse as animal body fluids/tissues or airborne dust, without prior culture (1).

   Bacterial carbohydrates make up a major portion of the cell wall, but a smaller proportion of the inner and outer (if present) membranes which are predominantly lipid. Mammalian cells do not possess a cell wall. Sugars found in mammalian cells are limited in variety. However, there is a great diversity of sugars found in bacteria, both common and rare (9,10). Our work involving identification and detection of unique bacterial sugar monomers has predominantly employed the alditol acetate method in conjunction with GC-MS using electron impact ionization. This work has successfully demonstrated that bacterial sugars can be readily identified in whole bacterial hydrolysates using total ion monitoring. The abundances of these sugars are often quite high, around 0.1-2.0% of the sample. In addition, sample size is not limited, allowing analysis of 5-10 mg of sample. The purpose of using selected ion monitoring is to improve visual discrimination of chromatograms by eliminating background and to increase sensitivity. However, it is possible to obtain mass spectra of sugars that are present at quite low concentrations (considerably less than 0.01%). Detection of bacteria in complex matrices is more difficult since bacterial markers are only present at trace levels. In this situation, selected ion monitoring is vital for detection of sugars present at low levels to distinguish the characteristic ion(s) amongst background noise.

The Alditol Acetate Method

The alditol acetate derivatization method for GC analysis of sugar mixtures was originally developed by Perry and co-workers (11) and popularized by Sawardecker for analysis of relatively simple samples such as purified glycoproteins (12). A number of methodological changes (including substantially improved clean-up steps and simplification of manual processing) have subsequently been made to allow ready identification of unique sugars in whole bacterial cells and for trace detection of bacterial remnants in complex matrices. Total ion analysis is required in this instance for sugar identification. Selected ion monitoring (SIM) removes minor contaminants from chromatograms of whole bacterial cell hydrolysates allowing ready quantitation. For trace detection of bacterial markers in complex matrices (e.g mammalian tissues or body fluids) SIM is essential.

A particular sugar of interest to us has been muramic acid, which is found in bacteria, but not elsewhere in nature (2, 13, 14). In the alditol acetate procedure neutral and amino sugars, such as muramic acid, are released by hydrolysis and then the aldehyde group is reduced (to eliminate the anomeric center) and hydroxyl and amino moieties subsequently acetylated. Elimination of the anomeric center simplifies chromatograms dramatically as most sugars produce one chromatographic peak. Muramic acid is exceptional in this regard in that it produces one major peak with a poorly resolved minor peak as a shoulder. The second peak has not been identified. Muramic acid produces a lactam (a cyclic amide) by internal dehydration between its carboxyl and amino groups. After acetylation, which involves extended high temperature heating, a tertiary amide is generated. Earlier versions of the alditol acetate procedure employed lower temperatures and shorter heating times; under these conditions, the carboxyl group of muramic acid does not react with the amino group. Thus, this compound will not pass through the GC system, due to absorption (15). Carboxyl groups, such as found in acidic sugars or amino acids, appear not to be derivatized under any of these reaction conditions and thus these compounds do not produce chromatographic peaks.

Between the reduction and acylation steps the reducing agent, (borohydride or borodeuteride) is removed (as tetramethylborate gas) by multiple evaporations with methanol/acetic acid which is quite time consuming. This is necessary since borate would inhibit the subsequent acylation reaction. The use of methylimidazole allows the acylation to proceed without removing the borate (16). However, extraneous peaks are generated from side-reactions between the catalyst (pyridine or methylimidazole) and acetic anhydride. Even with SIM this produces extremely dirty chromatograms. Thus, the use of methylimidazole is not recommended for trace analysis (17). As an alternative, an automated evaporator has been developed at USC for unattended removal of borate.

In addition to the selectivity achieved by using SIM, our procedure also incorporates pre-derivatization clean-up steps that remove hydrophobic substances (e.g. fatty acids) before acylation of hydroxyl/amino groups. Post-derivatization clean-up steps remove hydrophilic compounds (e.g acetylated amino acids that have free carboxyl groups). Thus, in the analysis of complex biological samples the procedure is highly selective for neutral and aminosugars that remain in the hydrophilic phase prior to derivatization, but in the hydrophobic phase post-derivatization.

Efforts to develop simpler methods to replace the alditol acetate procedure have not been successful. In the aldononitrile acetate procedure reaction of the aldehyde moiety with hydroxylamine (to destroy the anomeric center) replaces reduction with sodium borohydride; thus multiple evaporation steps to remove borate are eliminated. Extraneous peaks are generated from side-reactions between hydroxylamine and other derivatizing reagents such as acetic anhydride (13, 18). The introduction of active nitrile groups can produce irreversible adsorption on GC analysis particularly for certain aminosugars. Trimethylsilyl derivative of muramic acid and other sugars are simple to prepare but rather unstable (19, 20). Trifluoroacetates also rapidly decompose in the presence of moisture (21, 22). Furthermore, both of these procedures produce complex chromatograms (since the anomeric center is not eliminated and each sugar produces multiple peaks). For the aldononitrile acetate, trimethylsilyl and trifluoracetate methods clean-up procedures have not been optimized for profiling of neutral and amino sugars in whole bacterial cells or for trace detection in complex matrices.

Methodology

The alditol acetate procedure has been described in detail previously, including photographs of equipment used at each stage of the derivatization (15). This section, therefore, only briefly overviews the procedure, indicating significant changes.

Samples are hydrolyzed in 2N sulfuric acid under vacuum. Internal standards are added, followed by neutralization with N, N-dioctylmethylamine. Hydrophobic contaminants are removed by C-18 column extraction. During reduction, anomeric centers are often destroyed using sodium borohydride. Aldoses and alditols cannot be distinguished by this approach. Using sodium borodeuteride, asymmetry is retained, and these compounds can be distinguished. Methanol:acetic acid is then added to degrade sodium borohydride generating borate. During subsequent evaporations, borate is removed as tetramethylborate gas, which could otherwise complex with sugars, inhibiting the subsequent acylation.

Classically, borate removal involves five cycles of manual addition of methanol:acetic acid followed by evaporation. An automated evaporator has been custom built at USC. Samples are held in a turntable which sits in a heated water bath. Nitrogen is constantly passed into each sample through a manifold connected to a nitrogen tank. Samples rotate past a stationary central solvent reservoir from which methanol:acetic acid continuously flows (Figure 1A). For comparison, a conventional evaporator is shown in Figure 1B.

The samples are then heated under vacuum to remove any residual moisture prior to acylation. After cooling to room temperature, acetic anhydride is added and the samples are heated. Excess acetic anhydride is decomposed to acetic acid by the addition of water. Chloroform is added and the aqueous phase

Figure 1A. Equipment used in the preparation of batches of 21 mixtures of alditol acetates: automated evaporator in which samples are rotated past a reservoir of methanol:acetic acid while heated under nitrogen gas.

Figure 1B. Equipment used in the preparation of batches of 21 mixtures of alditol acetates: classical evaporation system employing heating of samples under nitrogen gas.

is discarded. Ammonium hydroxide is added followed by removal of the aqueous phase on magnesium sulfate columns.

After GC separation on an SP-2330 column, the MS detector is run in the total ion mode (for sugar identification) and selected ion mode (for quantitation). The resulting chromatograms provide a simple means for visual evaluation of major and minor components. For trace analysis samples are only analyzed in the SIM mode.

Mass Spectra of Alditol Acetates

A native sugar forms 2-4 anomers upon acylation thus creating multiple peaks from a single sugar and complicating chromatograms. Thus as noted above sugars are generally reduced using sodium borohydride or borodeuteride prior to acylation. For example, during reduction of aldoses the C-1 aldehyde is converted to an alcohol, (i.e. the aldose to an alditol) whereas alditols remain chemically unchanged. By using sodium borohydride in the formation of alditol acetates, aldoses and alditols can not be differentiated. However, when using sodium borodeuteride two deuteriums are added to the aldehyde moiety one of which remains after acylation. Thus the visual pattern (relative abundances of ions) of deuterated and non-deuterated mass spectra are similar except for a one mass unit shift in certain ions. Fragments lacking C-1 are not different between deuterated and non-deuterated samples. The mass spectra of fucose, xylose, and muramic acid will be discussed as examples.

The base peak in electron ionization (EI) mass spectra of alditol acetates is dominated by the acetylinium ion m/z 43. Many primary fragments are produced by cleavage between sequential carbon atoms. Secondary fragmentation results from losses of acetic acid (m/z 60), acetoxyl groups (m/z 59), and ketene (m/z 42) (23, 24).

The mass spectrum of fucose (mol. wt. 376), an example of an asymmetric molecule, is shown in Figure 2. Primary fragments in the non-deuterated spectra include m/z 303, m/z 289, m/z 231, and m/z 217. Examples of secondary fragments are m/z 201 (loss of 60 and 42 from 303), m/z 187 (loss of 60 and 42 from 289), m/z 170 (loss 59 and 60 from 289), and m/z 128 (42 from 170). Fragments which include C-1 change by 1 mass unit in the deuterated mass spectrum.

Xylose is an example of a symmetric molecule (see Figure 3). Thus ions of the same molecular weight can be generated from either end of the molecule. However fragments of deuterated molecules containing C-1 will be one mass unit greater than equivalent ions from the opposite end. Thus certain ions in non-deuterated mass spectra are represented as doublets in deuterated mass spectra. For example, m/z 115, 145, 187 and 217 are replaced by m/z 115/116, 145/146, 187/188 and 217/218 doublets in the deuterated spectrum. Alditols, which after reduction remain non-deuterated, will not exhibit these ion doublets. Thus simple visual observation of mass spectra, for the appearance of doublets allows one to readily distinguish aldoses from alditols. Primary and secondary fragments are produced as described for fucose.

Generally, amino sugar mass spectra are relatively simple since cleavage

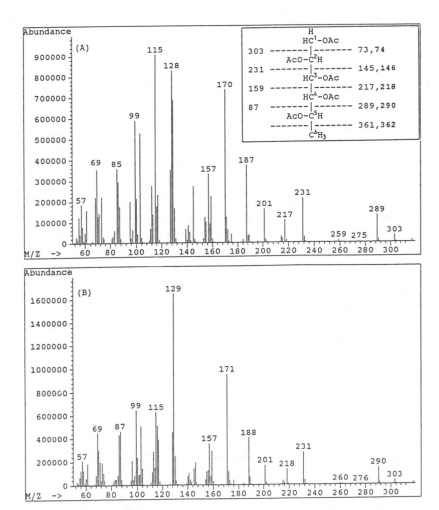

Figure 2. Mass spectra of alditol acetates of (A) Non-deuterated fucose (B) deuterated fucose

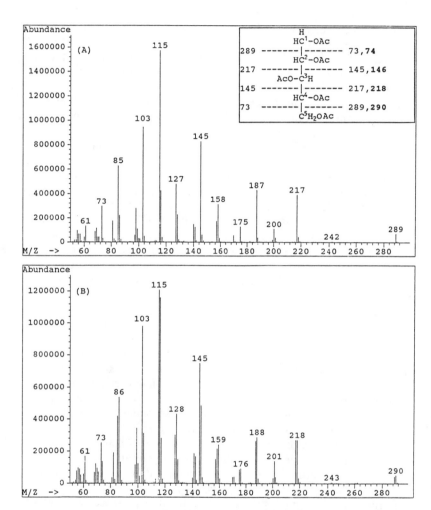

Figure 3. Mass spectra of alditol acetates of (A) Non-deuterated xylose (B) deuterated xylose

preferentially occurs between the carbon with attached acetamido group and adjacent acetylated carbons (25). The alditol acetate of muramic acid (3-O lactyl glucosamine) is more difficult to interpret, since it contains an internal, cyclic amide (see Figure 4). The molecular weight of muramicitol pentaacetate is 463. Loss of water allows the formation of an amide bond between the C-2 nitrogen and the carboxyl group of the lactyl moiety producing muramicitol pentaacetate lactam (molecular weight 445). The four hydroxyl moieties and the amino group appear to be acetylated (the latter existing as a tertiary amide) thus forming the pentaacetate. If the molecule contained a lactone, it would exist as a tetraacetate (2, 13, 26). It has been demonstrated by infrared spectroscopy that an acetate derivative of muramic acid exists as a lactam (13). Furthermore, muramic acid naturally exists as a lactam in spores (27). Carbons 4 through 6 extend as a straight chain arm from the ring and thus cleavage between C-3 of the ring and C-4 of the chain generates m/z 228. In general, alditol acetates do not form molecular ions under EI conditions, but presumably the ring form leads to stabilization of the molecular ion, m/z 445. Prominent secondary fragments include m/z 403 (ketene loss from 445), m/z 343 (60 from 403), m/z 283 (60 from 343), m/z 181 (60 and 42 from 283). A number of ions do not change in deuterated or non-deuterated spectra due to the loss of the deuterium containing C-1. This is evidenced by fragments m/z 312 (loss of 74 and 60 from 446), m/z 210 (loss of 60 and 42 from 312) and m/z 168 (loss of 42 from 210). Ion m/z 168 can also arise by loss of acetic acid from primary fragment m/z 228 (loss of the side chain). In deuterated samples, ion m/z 228 is replaced by ion m/z 229 since it contains C-1. In this case, the loss of acetic acid would yield ion m/z 169. Since m/z 168 can also arise from the fragments not containing C-1 (loss of 42 from 210) a doublet can be seen.

Naturally occurring O-methylated sugars are not common in bacterial cell wall polysaccharides. Although they have been described in eubacteria (28, 29) and in certain eukaryotes (30). However, O-methylated alditol acetates are commonly used in linkage analysis. As has been noted by a number of authors (24, 31), the fragmentation pattern of methylated sugars is distinctive. Fragmentation between the O-methylated carbon and the adjacent acetylated carbon atoms dominates the spectra. Additional secondary ions can be produced by loss of methanol (m/z 32) and formaldehyde (m/z 30).

The mass spectra of isomers of alditol acetates contain the same ions. It has been suggested that the small differences observed in relative intensities of these ions are insufficient for unambiguous differentiation of isomers. Furthermore, relative intensities of ion peaks have been noted to vary even for different runs with the same instrument (24). In our experience, however, certain isomers display differences in relative ion abundances which can be quite dramatic. We have previously noted that the aminodideoxyhexoses, quinovosamine and fucosamine (as found in legionellae), have quite distinct mass spectra (32).

The mass spectra of a 3-O methyl rhamnose standard, and 3-O methyl methylpentoses present in hydrolysates of B. anthracis and B. cereus have very similar relative ion abundances. In each case, for the pairs of ions m/z 88/101, m/z 130/143, and m/z 190/203 the lower molecular weight ion is in higher

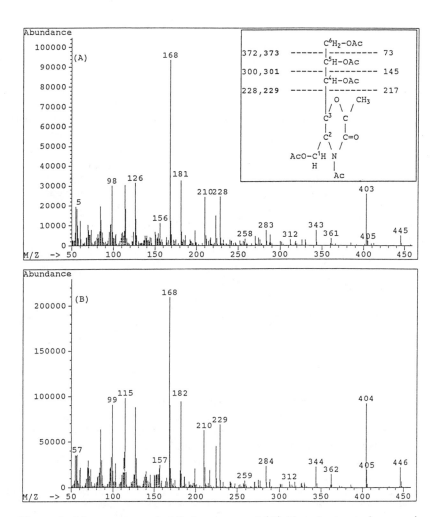

Figure 4.   Mass spectra of alditol acetates of (A) Non-deuterated muramic acid (B) Deuterated muramic acid

abundance. The opposite is seen for a 3-O methyl fucose standard. However, the closely related molecules 2-O methyl rhamnose and 2-O methyl fucose have fragmentation patterns which are not visually distinguishable (33).

The fragmentation pattern of many hexoses are also distinct. For example, glucose and galactose have similar mass spectra (see Figure 5). The galactose standard ion pair m/z 187/188 show essentially equal abundance. In the glucose standard, however, the m/z 187 peak predominates. Galactose from a hydrolysate of whole *B. anthracis* cells also displays essentially equal abundances of m/z 187/188.

Analysis of Whole Bacterial Cells by GC-MS

Bacterial cells contain sufficient amounts of carbohydrate that GC, combined with total ion EI-MS, can be used for identification of these sugars. For common sugars, GC retention time coupled with visual comparison of EI mass spectra allow ready identification. Bacterial species/genera often contain unusual sugars that are rarely found in nature. Standards are often unavailable for these compounds. In these instances, interpretation of EI mass spectra from first principles is essential. Positive ion chemical ionization allow production of a molecular ion which aids identification. With the availability of standards this is not usually necessary.

Classically, bacteria are identified by their physiological characteristics. Defining structural (chemical/molecular) composition is becoming an important alternative in modern taxonomic strategies (32). Examples of this include carbohydrate and fatty acid profiling, DNA-DNA hybridization, and ribosomal ribonucleic acid (rRNA) sequencing (34-36). The advantages of these approaches over conventional methods are their universal discriminating power, particularly when used in conjunction with one another. Chemical profiles and rRNA sequence each provide multiple features for bacterial differentiation. DNA-DNA hybridization and rRNA sequences, when used together, provide genetic information covering family, genus, and species levels. One should not ignore classical morphological/physiological discriminatory characteristics, but for many groups of bacteria, such as bacilli and legionellae, they are simply inadequate. A primary focus of research at USC is to develop profiling strategies that have widespread applicability among diverse taxonomic groups. Research has primarily focussed on two bacterial groups. The genus *Bacillus*, a Gram positive organism, and the *Legionellaceae*, a Gram negative family of organisms that contain at least two genera. In both instances there are numerous unresolved questions regarding the inter-relationships within and among the groups.

Much chemotaxonomic discrimination resides in the fatty acid or polysaccharide content of bacteria (37). With derivatization methods (for identifying chemical markers), depolymerization is usually achieved by acid hydrolysis, methanolysis, or saponification. Active functional groups on a monomeric compound are further derivatized (acetylated, esterified, etc.) to inert forms suitable for GC-MS. Coupling selective clean-up methods with derivatization enables simultaneous profiling of whole classes of compounds (e.g., sugars or fatty acids).

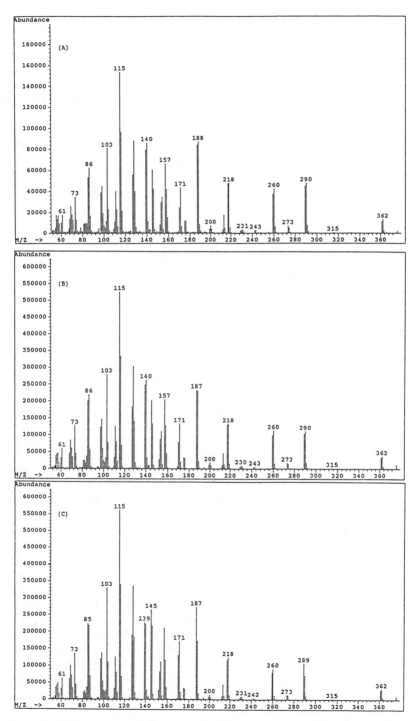

Figure 5. Mass spectra of deuterated alditol acetates of (A) galactose from *B. anthracis* (B) galactose standard (C) glucose standard.

Many chemicals commonly found in microorganisms are not well-suited for discrimination (e.g. glucose is found in most life forms). Some compounds (e.g. glucitol found in group B streptococci, or galactose (for differentiating *Bacillus anthracis* and *B. cereus*) are specific for various major groups of organisms and can serve as chemical markers for their detection or identification (1, 32).

Much of our earlier work has been concerned with a group of Gram negative bacteria - the *Legionellaceae*. Legionellae are important environmental agents that often colonize hot water towers. After airborne transmission, these microbes can initiate disease in susceptible individuals. *Legionella pneumophila* is the major pathogen in the family *Legionellaceae* and the causative agent of Legionnaire's disease. Because of their poor growth characteristics it has been difficult to use conventional biochemical tests in the differentiation of members of the *Legionellaceae*. We first differentiated legionellae by analysis of their carbohydrate content using gas chromatography with flame ionization detection (6). Total ion mode gas chromatography-mass spectrometry was used to detect a number of unusual sugars, including a branched octose (7) which was subsequently identified by others as yersiniose A (38). Increased sensitivity and selectivity for carbohydrate detection was achieved using selected ion monitoring (7). Two of the uncommon aminodideoxyhexoses we previously discovered in the legionellae (X1 and X2) were later identified as quinovosamine and fucosamine respectively (7, 39).

Recently, we have focussed on another group of environmental pathogens, the Gram positive bacilli. Many aspects of the taxonomic characterization and clinical identification of bacilli remain unresolved (40). The two major human pathogenic bacilli are *B. anthracis*, the causative agent of anthrax and *B. cereus*, a food-poisoning organism. *B. anthracis* has a high degree of genetic homology with *B. cereus* as demonstrated by DNA-DNA hybridization (41). Additionally the 16S rRNA sequences of these two species are almost identical (42) suggesting they may be part of a super-species. Physiologically and chemically these species are very similar. For example, identical cellular fatty acids are present in both species (43). Bacilli therefore represent an extreme test for the utility of carbohydrate profiling since many species are difficult to differentiate by both modern genetic and analytical chemical approaches. Furthermore, unlike most other groups of bacteria, bacilli exist in two distinct cellular forms (vegetative and spore).

B. anthracis and *B. cereus* have quite distinct carbohydrate profiles. Both contain ribose (presumably derived from RNA), glucose, muramic acid, glucosamine (derived from peptidoglycan) and mannosamine. For cultures in the vegetative form, *B. anthracis* is characterized by the presence of galactose, whereas *B. cereus* contains glucosamine. Spore forms of both organisms contain the characteristic sugars 3-O-methyl rhamnose and rhamnose. The spore form of *B. anthracis* also contains glucosamine. The spore form of *B. cereus* is quite distinct in that it additionally contains fucose and 2-0-methyl rhamnose. Naturally O-methylated sugars are uncommon but have previously been described among diverse groups of organisms, including eubacteria and eukaryotes (28-30). As far as we are aware O-methylated sugars have not previously been described in *B. anthracis* or *B. cereus*. Figure 6 illustrates the carbohydrate profiles of vegetative forms of *B. anthracis* and *B. cereus* (33).

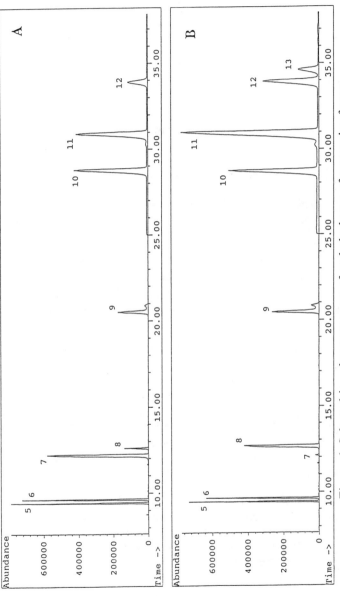

Figure 6. Selected ion chromatogram of carbohydrates of vegetative forms of (A) *B. anthracis* VNR-1-Δ1 and (B) *B. cereus* 1112. 5 = ribose, 6 = arabinose (internal standard), 7 = galactose, 8= glucose, 9 = muramic acid, 10 = methylglucamine (internal standard), 11 = glucosamine, 12 = mannosamine, 13 = galactosamine. (Reproduced with permission from ref. 33. Copyright 1993 American Society for Microbiology.)

To aid analyses, O-methylated rhamnose and fucose alditol acetate standards were synthesized. The mass spectra of O-methylated sugars are dominated by breakage of bonds between C-O-methyl and adjacent C-O-acetyl groups. For a deuterated 3-O-methyl methylpentose alditol acetate this would produce m/z 203 and m/z 190 as primary fragments (30). Secondary ions, as is the case for other alditol acetates, include loss of acetic acid (m/z 60) e.g. m/z 143 from m/z 203 and m/z 190 from m/z 130 or ketene (m/z 42) e.g m/z 101 from m/z 143 and m/z 88 from m/z 130. For a deuterated 2-O-methyl methylpentose this produces the ions m/z 118 and m/z 275 (28, 30).

GC, in conjunction with MS in the total ion mode, is a powerful tool for identification of unusual sugars in bacterial cells. Chromatograms produced are free of major contaminating peaks and are readily interpreted. Carbohydrate profiling of bacteria using selected ion monitoring removes minor peaks from chromatograms simplifying visual interpretation.

Trace GC-MS Analysis of Bacteria in Complex Matrices

Culture-based approaches provide only a rough measure of the amount of live bacteria present in complex matrices (such as mammalian tissues, body fluids, or airborne dust). Selective growth of certain species on culture often gives a false impression of the original population distribution. Remnants of disintegrated bacterial cells can function as well as viable bacteria as toxins. Measurement of viable bacteria can therefore provide false estimates concerning the bacterial contamination of different matrices. With suitable chemical methods, the total bacterial biomass is measured with no discrimination between live and dead organisms.

Processing of Bacterial Debris. There is a dearth of biochemical information on the events involved in the processing, dissemination and elimination of bacterial cell wall peptidoglycan (PG) in the mammalian host after infection. PG polymers, oligomers and monomers are all potent inflammogens. There are many factors involved in determining the severity and chronicity of inflammation elicited by PG. These include how rapidly PG is eliminated (in urine and via other routes) and how completely it is degraded by mammalian tissue enzymes (44).

Certain bacterial cell walls are highly resistant to degradation by mammalian enzymes, and thus can serve as a persisting inflammatory stimulus. The composition of peptidoglycan and associated polysaccharide (PS) affects the inflammatory properties of streptococcal cell walls. PG-PS is not rapidly eliminated from the host. The amount of certain antigenic determinants, including the terminal D-alanine dipeptide of PG as well as the rhamnose backbone and N-acetyl glucosamine side chains of PS, slowly decrease in amount *in vivo* (5, 45, 46). Rhamnose, muramic acid, and D-alanine are not synthesized by normal mammalian tissues. Over a 60 day period, the levels of PG-PS in the liver and joints of infected rats only decrease four-fold. This indicates that after an initial streptococcal infection in man there might be debris present almost indefinitely. Likewise, after an initial spirochete invasion of the joint, debris may persist for extensive periods in Lyme arthritis. Ocular inflammation can be elicited in rabbits by injection of muramyl dipeptide (MDP, a synthetic subunit of bacterial peptidoglycan). This model allows study of the etiology of the human disease, anterior uveitis. Muramic acid (derived from MDP) has been detected at concentrations of 1 part per ten million in rabbit serum 30 min after injection (14). However, within 4 h, this molecule is no longer detectable. The rapid elimination of MDP explains the self limiting nature of the inflammation. Normal rabbit serum, on the other hand, does not contain muramic acid.

The slow elimination of PG-PS from tissues could occur by several mechanisms. PG-PS egested or excreted (feces or urine) may contain intact or partially degraded forms. In addition to excretion, there may be some redistribution of PG-PS from the joint to depots in the reticuloendothelial system (RES). In some instances complete catabolic breakdown of PG-PS may occur. This would involve production of unique cell wall constituents which could then be converted into related mammalian substances. For example, one could speculate that rhamnose (once enzymatically released from PG-PS) might be converted into its isomer fucose. Although rhamnose is not a component of normal mammalian tissues, fucose is a component of mammalian glycoproteins. D-alanine, if formed as a breakdown product of PG-PS, might be converted into L-alanine and incorporated into mammalian protein. D-alanine is not found in mammalian tissues whereas L-alanine is a common amino acid found in proteins. Muramic might be converted into glucosamine by loss of its lactyl group. The catabolic pathways for *in vivo* degradation of bacterial debris have not been widely investigated.

Our work has suggested that muramic acid, one of the two major sugars in bacterial peptidoglycan, might be used as a generic marker for PG since it is not found elsewhere in nature (46, 47).

Chemical Markers for Bacteria in Organic Dust Present in Indoor Air. We are currently focussing our interests on chemically analyzing levels of airborne bacteria and cell wall components in organic dust. The organic composition of dust varies greatly depending on the environment. Dust may consist largely of animal matter (dust found in poultry houses) or plant matter (dust found in textile mills). In many instances, microbes or their constituents are natural contaminants in an environment. Several studies have found a correlation between the level of Gram negative bacterial endotoxin in indoor air and respiratory disease. "Sick buildings" have environments in which people experience certain clinical symptoms including itchy eyes and breathing difficulties. These symptoms have previously been ascribed to chemical contamination. More recently it has been suggested that bacterial contamination may also be involved. As an example, air conditioning systems are believed to be infected with bacteria. Currently, the only procedure for assessing the microbial concentration of indoor air is culture. The only widely used test for measuring endotoxin levels is the *Limulus* lysate test. There is a real need for an alternative standard approach for characterizing the microbial component of organic dust (3).

We are evaluating the utility of GC-MS to detect and monitor bacterial contamination in indoor air. The air monitored is from a variety of sources, including homes, hospitals, worksites and schools. Our primary focus is on the standardization of methods for characterizing the microbial content of air. Culture-based approaches provide only a rough measure of the amounts of live bacteria present in air. Because culture can result in selective growth, a false impression of bacterial population distribution is often seen. In addition, remnants of disintegrated, bacterial cells can equally well function as allergens as can viable bacteria. Thus measurements of the levels of viable bacteria can provide false estimates of the allergenic content of air. With chemical methods, the total bacterial load is measured with no discrimination between live and dead organisms.

There has been a major focus on bacterial endotoxins (lipopolysaccharide, LPS) derived from Gram negative bacteria. Traditionally, levels of LPS (present in live or dead bacteria) have been measured by the *Limulus* lysate assay. Although useful, this assay has been demonstrated to have a rather limited specificity. Other bacterial toxins, including peptidoglycan (PG) and non-microbial derived chemicals can activate the *Limulus* lysate enzyme cascade. Certain hydroxy fatty acids can be used as chemical markers for more accurately assessing the levels of endotoxin in contaminated air. The technique utilized has exquisite sensitivity (negative ion GC-MS analysis of halogenated derivatives) (3, 4, 48).

In certain indoor air environments (e.g. poultry and swine farms), Gram positive bacteria may greatly outnumber Gram negative bacteria. In these instances, it would be useful to have a measure of Gram positive toxins. PG is found in almost all bacterial species and is known to produce many endotoxic effects. PG is also capable of causing chronic inflammatory reactions (44). A marker widely distributed among bacterial species but absent elsewhere in nature is muramic acid (found in the bacterial cell wall PG).

In some instances, pulmonary problems can result from infection with specific environmentally derived airborne pathogens. The most well-known of these is perhaps *Legionella pneumophila*, the causative agent of Legionnaires disease. It is extremely difficult to quantitatively evaluate the levels of certain airborne respiratory pathogens (e.g. legionellae and mycobacteria) as they grow poorly in conventional microbiological media. Concentrations of species or genus specific marker compounds can be monitored directly to provide a non-culture based quantitative measure of these species in the air. Specific examples of these are the aminodideoxyhexoses as found in most legionellae and tuberculostearic acid present in mycobacteria (6, 7, 49).

For analysis of air-borne bacterial contamination, dust was collected from the air conditioning filters of a number of homes and hospitals (50). In most homes the indoor environment is directly connected, via a single air conditioning air filter, to the outside environment. Because of concern about opportunistic bacterial infections, hospital filtration systems are more elaborate and designed to remove as much dust and associated bacterial particles as possible. Glucose was added (1 mg/mL) to all samples to serve as a carrier for muramic acid. The addition of 2 μg of methylglucamine as an internal standard prior to GC-MS allowed quantification. Initial analyses consisted of standards of muramic acid (30 ng to 18 μg), carrier and internal standard run in parallel with negative controls (no muramic acid). In order to increase sensitivity the samples were dissolved in the minimal sample volume for the autosampler (30 μl) and a 5 μl aliquot analyzed under non-purging conditions. MS monitoring was set at "low resolution mode" to further increase sensitivity. Muramic acid at 30 ng was readily detected with no muramic acid peaks in companion blanks. Figure 7 demonstrates the presence of muramic acid in dust collected from a home filter (approximately 20 mg dust). Selected ion monitoring of ions m/z 168, 404, and 446 (each characteristic of muramic acid), only detect one major peak at the correct retention time. The relative abundance of these ions was also appropriate. The levels of muramic in homes and hospitals ranged from 2-50 ng per mg of dry dust. The large tailing peak seen eluting before muramic acid is derived from the carrier glucose. Ion m/z 170 was monitored to detect the internal standard methylglucamine. Glucosamine, which is structurally related to muramic acid, was also detected in all samples. This sugar occurs widely throughout nature in higher animals, plants, fungi and bacteria.

In summary, bacteria and bacterial debris can be detected in airborne dust without prior culture by searching for unique chemical markers. The analysis of such trace components is considerably more complex than analysis of a monomer in isolated bacterial cells. Greater selectivity in derivatization and clean-up procedures is required as well as the selection of appropriate MS instrumental conditions. A number of unique compounds found in microbes are of potential utility as chemical markers. These include muramic acid (found in most pathogenic bacteria), β hydroxymyristic acid, and L-glycero-D-mannoheptose, (components of Gram-negative lipopolysaccharide). Certain sugars, such as aminodideoxyhexoses, have a much more limited distribution in nature and might be used to monitor the presence of airborne pathogens such as legionellae. The detection of chemical markers for microbes eliminates time consuming culture and could dramatically improve the determination of the microbial content of airborne dust.

Future Directions

Over the past fifteen years we have continued to optimize the alditol acetate procedure for GC-MS analysis. A fully automated procedure is a realistic possibility thanks to the availability of workstations and robotics which can replace many aspects of manual sample handling. These include solvent addition, heating, column-cleanup and evaporation. Application of the newly developed automated evaporator for borate removal has greatly simplified the existing method. Current limitations of this procedure are the amount of time required for derivatization and the need for increased sensitivity. The procedure as currently performed allows both identification of bacteria and their detection in complex matrices. Alternatives for these two areas need to be considered independently since the major issues to be resolved in each are quite different.

Identification of Bacteria. Many bacterial chemical markers are present in samples at quite high abundance. Additionally, many bacteria can be readily grown up in large amounts. Therefore, provided the MS procedure has adequate sensitivity, the preferred approach would be the one that requires the least sample handling. Developments in structure analysis of underivatized polar compounds

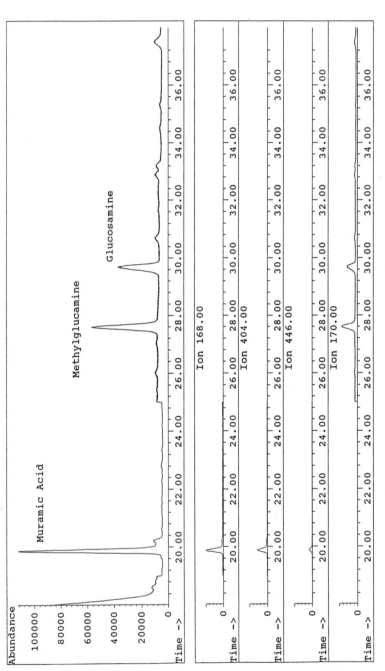

Figure 7. Selected ion chromatogram of home dust collected from an air conditioning filter demonstrating the presence of muramic acid. Ion m/z 168, 404 and 446 each detect muramic acid at the correct retention time. Ion 170 detects the internal standard methylglucamine.

by mass spectrometry are proceeding rapidly (51, 52). After release from whole bacterial cells, sugar and fatty acid markers can be analyzed by liquid chromatography-mass spectrometry (LC-MS) (53, 54) or tandem MS (MS-MS) (55, 56). These recent developments have been possible particularly due to the availability of new ionization techniques (51). The sensitivity of LC-MS (of underivatized sugars) is not as high as GC-MS (of derivatized samples) although routine profiling of whole bacterial cells appears to be feasible. MS-MS cannot readily distinguish isomers, particularly if more than one isomer is present in a sample. Many distinctions in bacterial sugar profiles involve the differentiation of isomers, which are well resolved by GC or LC. Further developments in LC-MS and MS-MS may allow simple and rapid bacterial identification.

Trace Detection of Bacteria in Complex Matrices. There is great potential for the use of MS to assess the levels of bacterial contamination in various situations. For example, determining the purity of biotechnology and pharmaceutical products, monitoring air and water for levels of bacterial contamination, and assessing levels of bacteria in infectious diseases. Bacterial products are often potent even at extremely low concentrations, often nanograms or picograms per mL, thus maximizing sensitivity is essential. For trace analysis of chemical markers in more complex matrices, halogenated derivatives can be prepared, separated by GC, and detected by negative ion chemical ionization (NI-CI) MS (3, 49, 58). There are also exciting possibilities for GC-MS-MS and NI-CI for microbial detection (57). The sensitivity of NI-CI MS with halogenated derivatives of bacterial chemical markers is much greater than conventional detection. Fatty acids (eg. tuberculostearic and β hydroxymyristic acids) (3, 49), amino acids (eg. diaminopimelic acid and D-alanine) (4, 5), as well as sugars (eg. muramic acid) (58, 59), have been successfully analyzed by NI-CI.

In the development of analytical methods for trace detection of a chemical marker, it may be useful to consider sensitivity and selectivity as distinct issues. Sensitivity refers to the absolute level at which a pure substance can be detected. Selectivity refers to discrimination among a complex mixture of compounds, particularly when the compound of interest is present at a much lower concentration. If a procedure lacks either sensitivity or selectivity, the lower limits of detection are adversely affected.

Among the most selective and sensitive GC (or LC) detectors is the MS. Electron impact (EI) ionization for detection of positive ions is quite selective and sensitive. Use of negative ion chemical ionization of halogenated derivatives increases both selectivity and sensitivity. Underivatized material is not detected due to a lack of electron capturing groups. Minimal fragmentation occurs due to the gentler ionization reaction thus ensuring production of a limited number of ions (predominantly molecular ions) and decreasing background noise. Halogenated derivatives are extremely strong electron capturing compounds so sensitivity is also greatly enhanced. The selectivity of MS based detection can also be improved by efficient clean-up of samples prior to GC-MS analysis.

EI-MS has been used with great success for the detection of bacterial sugars (including muramic acid) in complex matrices. The alditol acetate procedure produces simple chromatograms which are free of extraneous background noise. Muramic acid, as a component of muramyl dipeptide (a synthetic subunit of bacterial peptidoglycan), has been detected at concentrations of 100 ng/mL in rabbit serum (1 part per ten million) (14).

A new procedure for the trace analysis of muramic acid, consisting of methanolysis followed by trifluoroacetylation, was recently developed (58). This current method was adapted from the procedure of Pritchard, Coligan, Speed and Gray (19), which has been used successfully for electron capture detection but not previously used in conjunction with NI-CI detection. In addition to producing methyl glycosides from hydroxyl groups (which are involved in linkage of monomeric units), the carboxyl group of muramic acid is methylated. Subsequently, free hydroxyl and amino groups are trifluoroacetylated. In order to reduce the number of manual steps in the procedure, a single clean-up to remove lipids after methanolysis was employed. Clean-up steps occurring after acylation are difficult to incorporate since TFA derivatives are often unstable in the presence of moisture. The selectivity and sensitivity of this procedure is achieved employing NI-CI detection.

There have been earlier efforts to prepare halogenated derivatives suitable for NI-CI analysis for muramic acid, but these methods have not been widely used (59, 60). The methyl TFA procedure is at least 100 times more sensitive than the alditol acetate procedure. In practical terms, NI-CI analysis of the methyl TFA procedure requires the use of far less starting material. When sample is limited this is a distinct advantage of the methyl TFA procedure over the alditol acetate procedure. If a sample is not limited in amount, both methyl TFA and alditol acetate procedures can detect comparable concentrations of muramic acid in complex samples. The selectivity of chemical ionization, followed by negative ion detection allows the elimination of several clean-up steps as contaminants are often not ionizable. Muramic acid produces a single major peak with a poorly resolved minor component. Other sugars (methyl glycosides/TFA derivatives) produce as many as four well resolved peaks under these conditions. Total ion chromatograms for mixtures of neutral and aminosugars from whole bacterial cells become quite complex. Therefore, bacterial carbohydrate profiles are best analyzed as alditol acetates.

To this point the only widely used commercial halogenated derivatizing reagents are trifluoroacetyl, pentafluoropropyl and heptafluorobutyl anhydride. All have been used successfully in preparing halogenated carbohydrate derivatives from bacteria (19, 59, 60). Unfortunately, all of these halogenated derivatives of sugars are unstable in the presence of moisture. Post-derivatization clean-up steps are therefore not possible. This limits the selectivity (and thus limits detection) of trace components in complex mixtures since the halogenated derivatizing agent can not be removed (producing background noise). There is a real need for the development of a new procedure for the generation of stable halogenated sugar derivatives.

Conclusions

The alditol acetate procedure is a reliable method for identification of novel sugars in whole bacterial cells allowing chemotaxonomic differentiation. Trace detection in complex matrices of bacteria or their products is also readily achievable. Further automation and simplification of the alditol acetate procedure will encourage widespread application of the method.

Acknowledgements

This work was supported by the Army Research Office and the Center for Indoor Air Research. Gavin E. Black was funded by a training award from the Army Research Office DoD EPSCoR program. The authors wish to thank Mike Gore for his help in building the prototype automated evaporator.

Literature Cited

1. *Analytical microbiology methods: chromatography and mass spectrometry*; Fox, A.; Larsson, L.; Morgan, S.; Odham, G., Eds.; Plenum: NY, NY, 1990.

2. Fox, A.; Schwab, J.H.; Cochran, T. *Infect. Immun.* **1980**, *29*, 526.

3. Sonesson, A.; Larsson, L.; Westerdahl, G.; Odham, G. *J. Chromatogr.* **1987**, *417*, 11.

4. Sonesson, A.; Larsson, L.; Fox, A.; Westerdahl, G.; Odham, G. *J. Chromatogr.* **1988**, *431*, 1.

5. Ueda, K.; Morgan, S.L.; Fox, A.; Gilbart, J.; Sonesson, A.; Larsson, L.; Odham, G. *Analyt. Chem.* **1989**, *61*, 265.

6. Fox, A.; Lau, P.Y.; Brown, A; Morgan, S.L.; Zhu, Z-T; Lema, M; Walla, M.D. *J. Clin. Micro.* **1984**, *19*, 326.

7. Fox, A.; Rogers, J.C.; Fox, K.F.; Schnitzer, G.; Morgan, S.L.; Brown, A.; Aono, R. *J. Clin. Microbiol.* **1990**, *28*, 546.

8. Roboz, J.; Suzuki, R.; Holland, J.F. *J. Clin. Microbiol.* **1980**, *12*, 594.

9. Lindberg, B. *Adv. Carb. Chem. Biochem.* **1990**, *48*, 279.

10. Mayer, H.; Tharanathan, R.N.; Weckesser, J. In *Methods in Microbiology*; Gottschalk, G., Ed; Academic Press: NY, NY, 1985, Vol. 18; 157-208.

11. Gunner, S.W.; Jones, J.K.N.; Perry, M.B. *Chem. Ind.(London)*, **1961**, 255.

12. Sawardeker, J.S.; Sloneker, J.H.; Jeanes, A. *Anal.Chem.* **1965**, *37*, 1602.

13. Findlay, R.H.; Moriarty, D.J.W.; White, D.C. *Geomicrobiol. J.* **1983**, *3*, 135.

14. Fox, A.; Fox, K. *Infect. Immun.* **1991**, *59*, 1202.

15. Fox, A.; Morgan, S.L.; Gilbart, J. In *Analysis of carbohydrates by GLC and MS*; Biermann, C.J.; McGinnis, G.D., Eds.; CRC Press: Boca Raton, FL, 1989; 87-117.

16. Chen, C.C.; McGinnis, G.D. *Carbohydr. Res.* **1981**, *90*, 127.

17. Whiton, R.S.; Lau, P.; Morgan, S.L.; Gilbart, J.; Fox, A. *J. Chromatogr.* **1985**, *347*, 109.

18. Eudy, L.W.; Walla, M.D.; Morgan, S.L.; Fox, A. *Analyst* **1985**, *110*, 381.

19. Pritchard, D.G.; Settine, R.L.; Bennet, J.C. *Arth. Rheum.* **1980**, *23*, 608.

20. Kakeki, K.; Honda, S. In *Analysis of carbohydrates by GLC and MS*; Biermann, C.J.; McGinnis, G.D., Eds.; CRC Press: Boca Raton, FL, 1989; 43-85.

21. Pritchard, D.G.; Colligan, J.E.; Speed, S.F., Gray, B.M. *J. Clin Microbiol.* **1981**, *13*, 89.

22. Bryn, K.; Jantzen, E. *J. Chromatogr.* **1982**, *240*, 405.
23. Biemann, K.; DeJongh, D.C.; Schnoes, H.K. *J. Am. Chem. Soc.* **1963**, *85*, 1763.
24. Lindberg, B. In *Methods in Enzymology*; Ginsburg,V., Ed.; Academic Press: NY, NY, 1972, Vol. 28; 178-194.
25. Longren, J.; Svenssen, S. *Adv. Carbohydr. Chem. Biochem.* **1974**, *29*, 41.
26. Walla, M.D.; Lau, P.Y.; Morgan, S.L.; Fox, A.; Brown, A. *J. Chromatogr.* **1984**, *288*, 399.
27. Warth, A.D.; Strominger, J.L. *Biochem.* **1972**, *11*, 1389.
28. Brennen, P.J.; Mayer, H.; Aspinall, G.O.; Nam Shin, J.E. *Eur. J. Biochem.* **1981**, *115*, 7.
29. Saddler, G.S.; Tavecchi, P.; Lociuro, S.; Zanol, M.; Colombo, L.; Selva, E. *J. Microbiol. Meth.* **1991**, *14*, 185.
30. Talmont, F.; Fournet, B. *FEBS Lett.* **1991**, *281*, 55.
31. Carpita, N.C.; Shea, E.M. In *Analysis of carbohydrates by GLC and MS*; Biermann, C.J.; McGinnis, G.D., Eds.; CRC Press: Boca Raton, FL, 1989; 87-117.
32. Morgan, S.L.; Fox, A.; Rogers, J.C.; Watt, B.E. In *Modern techniques for rapid microbiological analysis*; Nelson, W.H., Ed.; VCH Publishers: NY, NY, 1991; 1-18.
33. Fox, A.; Black, G.E.; Fox, K.; Rostovtseva, R. *J. Clin. Micro.* **1993**, *31*, 887.
34. Woese, C.R. *Microbiol. Reviews* **1987**, *51*, 221.
35. Olsen, G.J.; Lane, D.J.; Giovannoni, S.J.; Pace, N.R. *Annu. Rev. Microbiol.* **1986**, *40*, 337.
36. Schleifer, K.H.; Stackebrandt, E. *Ann. Rev. Microbiol.* **1983**, *37*, 143.
37. Moss, C. W. In *Analytical Microbiology Methods: chromatography and mass spectrometry*; Fox, A.; Larsson, L.; Morgan, S.L.; Odham, G., Eds.; Plenum Press: NY, NY, 1990; 59-69.
38. Sonesson, A.; Jantzen, E. *J. Microbiol. Meth.* **1992**, *15*, 241.
39. Sonesson, A.; Jantzen, E.; Bryn, K.; Larsson, L. *Arch. Microbiol.* **1989**, *153*, 72.
40. Claus, D.; Fritze, D. In: *Bacillus*; Harwood, C.R., Ed.; Biotechnology Handbooks, v.2; Plenum Press: NY, NY, 1989; 5-26.
41. Kaneko, T.; Nozaki, R.; Aizawa, K. *Microbiol. Immunol.* **1978**, *22*, 639.
42. Ash, C.; Farrow, J.A.E.; Dorsch, M.; Stackebrandt, E.; Collins, M.D. *Int. J. System Appl.* **1991**, *41*, 343.
43. Lawrence, D.; Heitefuss, S.; Seifert, H.S.H. *J. Clin. Microbiol.* **1991**, *29*, 1508.
44. Fox, A. *Acta Path. Micro. Immunol. Scand.* **1990**, *98*, 957.
45. Eisenberg, R.; Fox, A.; Greenblatt, J.J.; Anderle, S.K.; Cromartie, W.J.; Schwab, J.H. *Infect..Immun.* **1982**, *38*, 127.
46. Gilbart, J.; Fox, A. *Infect. Immun.* **1987**, *55*, 1526.
47. Gilbart, J.; Fox, A.; Whiton, R.S.; Morgan, S.L. *J. Microbiol. Meth.* **1986**, *5*, 271.
48. Mielniczuk, Z.; Alugupalli, S.; Mielniczuk, E.; Larsson, L. *J. Chrom.* **1992**, *623*, 115.
49. Larsson, L.; Odham, G.; Westerdahl, G.; Olsson, B. *J. Clin. Microbiol.* **1987**, *25*, 893.

50. Fox, A.; Rosario, R.M.T. Proceedings of Indoor Air '93 [6th international conference on indoor air quality and climate]. (in press)

51. *Methods in Enzymology*; McCloskey, J., Ed.; Academic Press: NY, NY, 1990, Vol. 193, 1-960.

52. Lin, H-Y; Voyksner, R.D. *Anal. Chem.* **1993**, *65*, 451.

53. Simpson, R.C; Fenselau, C.C.; Hardy, M.R.; Townsend, R.R.; Lee, Y.C.; Cotter, R.J. *Anal. Chem.* **1990**, *62*, 248.

54. Elmroth, I.; Larsson, L.; Westerdahl, G.; Odham, G. *J. Chrom.* **1992**, *598*, 43.

55. Cole, M.J.; Enke, C.C. *Anal. Chem.* **1991**, *63*, 1032.

56. Platt, J.A.; Uy, O.M.; Heller, D.N.; Cotter, R.J.; Fenselau, C. *Anal. Chem.* **1988**, *60*, 1415.

57. Johnson, J.V.; Yost, R. *Analyt. Chem.* **1985**, *57*, 758A.

58. Elmroth, I.; Fox, A.; Larsson, L. *J. Chromatogr.* **1993**, *628*, 93.

59. Christensson, B.; Gilbart, J.; Fox, A.; Morgan, S.L. *Arth. Rheum.* **1989**, *32*, 1268.

60. Tunlid, A.; Odham, G. *J. Microbiol. Meth.* **1983**, *1*, 63.

RECEIVED June 7, 1993

Chapter 9

Mass Spectrometric Determination of D- to L-Arabinitol Ratios for Diagnosis and Monitoring of Disseminated Candidiasis

J. Roboz

Department of Neoplastic Diseases, Mount Sinai School of Medicine,
New York, NY 10029

A review of candidiasis as well as an analytical method for its detection are described. The methodology can be applied to easily collected samples of unmeasured drops of whole blood, serum, or urine spots on mailable filter paper. Enantiomers of arabinitol are separated by chiral chromatography and analyzed using chemical ionization and selected ion monitoring. Because only D/L arabinitol ratios are determined, no quantitative calibration is needed. The method is fast: ten samples/day, single samples in two hours. Analyzing both serum and urine may increase reliability. D/L>3.3 (normal mean+3 st.dev.) in serum or >2.5 in urine suggests candidiasis. Diagnostic sensitivity was 95% (1/22), specificity 88% (43/49) in an uninfected population heavily biased (50%) to renal dysfunction. Suggested applications: aiding diagnosis, monitoring patients at risk so that treatment can be initiated while fungus load is still small, and following therapeutic or prophylactic antifungal chemotherapy.

Disseminated candidiasis in immunosuppressed patients is a devastating disease with high morbidity and mortality. Because of the protean presentation of this infection, and the unreliability of diagnosis by blood culturing or by serological techniques, antemortem diagnosis is difficult and treatment is often unsuccessful. After reviewing the clinical and diagnostic literature on this disease, this chapter summarizes recent advances in the mass spectrometric determination of D/L arabinitol ratios in blood and urine and some clinical applications for the diagnosis and monitoring of disseminated candidiasis.

Candidiasis

Opportunistic Infection. Recent advances in the early diagnosis and improved treatment of cancer has produced cures in nearly half of all cancer patients, adults as well as children. The consequence of aggressive cytotoxic, corticosteroid or radiotherapies is frequently the induction of profound

0097–6156/94/0541–0132$06.00/0

neutropenia (neutrophil count < 1,000/mL) which places the patients at high risk of complications due to opportunistic infections (OI). The longer the patient is neutropenic, the more likely OI occurs. Other patients at risk of OI include those who have received organ transplants, have undergone extensive surgery (e.g., abdominal surgery for solid tumors), or treated for burns. Although bacterial infections are the most frequent OI, the availability of new antibacterial agents of increasingly higher potency and broader spectrum has led to an increased frequency of fungal OI over the past decades *(1)*. Candidiasis accounts for the majority of serious fungal infections.

Candida **organisms.** *Candida* organisms (small ovoid yeasts, 4-8 um diameter) are ubiquitous; they are normal commensals in the oropharynx, intestinal tract and vagina. When conditions are propitious, overgrowth, tissue invasion, and dissemination into deep-seated organs (kidney, lung, liver, heart) leads to candidiasis (known as candidosis in Europe), defined as a local or systemic infection caused by *Candida*. At least eight species of the genus *Candida* have been found to be pathogenic for man. Of these *C. albicans* is predominant (order of 50%) and, together with *C. glabrata* (formerly *Torulopsis glabrata*), is prevalent in solid tumors, while *C. tropicalis* or *C. krusei* are more common in patients with hematologic malignancies *(2)*.

Clinical Manifestations. The clinical spectrum of candidiasis is very broad, ranging from mucocutaneous candidiasis, through colonization of the alimentary and urinary tracts, to candidemia and disseminated candidiasis. Superficial candidiasis of the skin and mucous membranes is easily diagnosed clinically and responds readily to treatment in the non-immunocompromised patient. The term systemic candidiasis is general, including candidemia (the presence of culturable organisms in peripheral blood), disseminated candidiasis (spreading from a focus at a mucous surface or in the skin to deep-seated organs), and infection of sites of non-hematogenous origin. Patients with neutropenia are prone to life-threatening systemic candidiasis, often by invasion through the gastrointestinal tract or by way of indwelling intravenous catheters or parenteral feeding tubes. Patients with defects of cell-mediated immunity (including those infected with the human immunodeficiency virus) are more likely to develop mucocutaneous candidiasis.

Epidemiology. A four year study on 233 patients revealed that *Candida* species caused 74%, 61% and 92% of fungal infections in patients with acute leukemia, lymphoma, and solid tumors, respectively *(3)*. The incidence of disseminated candidiasis in an autopsy series of children with various cancers was 10% with 88% having more than one deep organ affected *(4)*. *Candida* species accounted for 79% of nosocomial fungi in the past decade *(5)*. Disseminated candidiasis has been found on the increase among children infected with the HIV virus *(6)*.

Treatment. Disseminated candidiasis presents both diagnostic and atherapeutic challenges *(7)*. Because of the lack of diagnostic procedures, the majority of cases are not recognized sufficiently early to permit effective treatment. The main therapeutic agent remains Amphotericin B. Due to the toxicity of systemic antifungal therapy, treatment is often undertaken with considerable reluctance in the absence of definitive diagnosis. Nevertheless, there is a strong recent trend for empirical antifungal therapy and prevention in persistently febrile neutropenic patients; current approaches include new antifungal agents such as fluconazole and lyposomal Amphotericin B, and the immunomodulators such as granulocyte colony-stimulating factor, and granulocyte-macrophage colony-stimulating factor *(8-11)*.

The study of candidiasis involves a wide range of disciplines, both in the basic and clinical sciences, leading to a large number of publications on the subject. A major book, published in 1988, lists nearly 6000 papers *(12)*; more than two thousand more have been published since.

Diagnosis

Signs and Symptoms. The diagnosis of oral candidiasis, the most common form of superficial candidiasis, generally is not difficult. Thrush lesions (superficial white elevated plaques that resemble milk curds) can readily be identified and confirmed by microscopy. Although there are detailed descriptions of the signs and symptoms of the many types of candidiasis *(13)*, the regretful fact is that the clinical presentation of invasive candidiasis is protean, there are no distinctive clinical manifestations, and often the clinical features are overshadowed by symptoms of the underlying disease. Accordingly, presumptive clinical diagnosis is usually based on fever of unclear etiology, macronodular skin lesions, endophtalmitis, and diffuse severe tenderness *(14)*.

Definitive Diagnosis. The only definitive method of diagnosis of disseminated candidiasis is the direct demonstration of typical morphological forms in tissues obtained by biopsy, to be confirmed by the culture of the biopsy. Obviously, this approach is often contraindicated when deep-seated organs are involved. Diagnostic imaging techniques, e.g., CT scanning and magnetic resonance imaging are inadequate for early diagnosis.

Blood Culturing. Although widely used, blood cultures are not reliable *(3,15)*. According to Meunier: "Patients who are colonized do not necessarily have an invasive infection. On the other hand, 30 to 50% of patients with autopsy-proven infection lack relevant cultures pre-mortem" *(16)*.

Serodiagnosis. More recent approaches to the serodiagnosis of disseminated candidiasis include those based on the detection of antibodies to *Candida*, detection of circulating candidal antigens, and quantification of circulating characteristic metabolites or cell wall constituents of the organism. Reviews of these techniques emphasize that despite of impressive technological progress, most techniques lack adequate sensitivity and specificity *(17-19)*. Recent techniques, e.g., the *Candida* antigen latex *(20,21)*, and circulating *Candida* enolase *(22)* tests are involved and not yet ready for routine use.

D-Arabinitol as Candida Metabolite

Identification. In 1979, Kiehn and coworkers identified D-arabinitol, a pentitol, as a major metabolite of several pathogenic *Candida* species, in the organisms *(23)* and in experimental animals and patients with confirmed disseminated candidiasis *(24)*. The identity of arabinitol was confirmed by obtaining the chemical ionization (CI) mass spectra of both the trimethylsilyl (TMS) and peracetate derivatives. The fact that the arabinitol produced by *Candida* is of the D enantiomer was confirmed by melting point measurements using the crystallized peracetate derivative. Wong et al. also developed a gas chromatographic (GC) technique (using the TMS derivative and flame ionization detection) for quantification of arabinitol in serum and urine, and reported data from normals, patients, and infected experimental animals *(25)*. Additional mass spectrometric confirmation of the identity of arabinitol as the marker in both culture and serum was reported by Roboz et al. using high resolution for the accurate mass measurement of the protonated molecule *(26)*. They determined that virtually 100% of the arabinitol produced by *C. albicans* is of the D form *(27)* by gas chromatography-mass spectrometry (GC/MS), using chiral separation of the arabinitol enantiomers.

"Arabinitol" and Total Pentitols. As shown in Figure 1, there are three pentitols, ribitol (adonitol), xylitol, and arabinitol (arabitol); only arabinitol has D and L enantiomeric forms. All earlier techniques for the quantification of arabinitol, both GC and GC/MS, used packed chromatographic columns which cannot separate the three pentitols. Thus, most "arabinitol" concentrations reported in the literature represent, in fact, the sum of the three pentitols, including both arabinitol enantiomers. As discussed later, there are considerable differences in the diagnostic utility of data based on total pentitol, total arabinitol, and separated D- and L-arabinitol determinations.

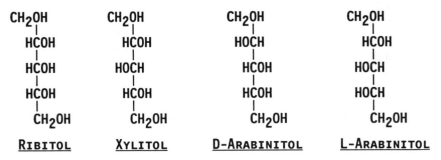

Figure 1. Formulas of pentitols

GC/MS Quantification Techniques

Total Pentitols. Table I lists, in chronological order, the GC/MS methodologies which have been developed. The first procedure *(26)* , developed for serum, separated trimethylsilylated arabinitol usingGC columns (1 m long, 2 mm diameter) packed with OV-7 (3%, methyl-phenyl silicone gum) which were used isothermally at 175-200 °C range. Trapped endogenous serum constituents were periodically removed at 250 °C. Starting with 0.2 mL serum, proteins were precipitated with acetone, the supernatant was evaporated to dryness and the residue derivatized using N,O-(bis)trimethylsilyltrifluoroacetamide+1% trimethylchlorosilane (BSTFA+1% TMCS, 3:1 v/v)) at 100 °C for 5 min. Because of frequent interferences, both erythritol and 2-deoxygalactitol were initially used as internal standards; in subsequent work the latter was used preferentially because of its more convenient chromatographic properties. The TMS derivative of arabinitol was detected and quantified by GC/MS by selected ion monitoring of the protonated molecules of arabinitol (m/z=513), 2-deoxy-galactitol (m/z=527) and erythritol (m/z=411) obtained by positive CI with isobutane as both the chromatographic carrier gas and the CI reagent gas. Because of the ever present endogenous arabinitol, quantification was carried out using the standard addition method. Calibration curves were obtained by adding increasing quantities of arabinitol to a normal serum sample (keeping the concentration of int. std. constant); extrapolation of the calibration curve to the abscissa provided the amount of the endogenous arabinitol.

Table I. Evolution of GC/MS Methodologies

Objective	GC column	Derivative	MS mode	Reference
Total Pentitols: D+L arabinitol + ribitol + xylitol	Packed	TMS	+ CI	26
Individual polyols Total arabinitol: D+L	Packed/ Capillary	TFA	+ CI	28
Individual pentitols incl. total arabinitol	Capillary	TMS	+ CI	29
Separated D- and L-arabinitol D/L ratio or quantify	Chiral	TFA	+ CI	27
Filter paper spots of whole blood, urine D/L ratio or quantify	Chiral	TFA	- CI	31, 32

Individual Polyols. With the advent of capillary GC columns, separation of the individual polyols (but not of the arabinitol enantiomers) became relatively easy. A GC/MS method was developed for the separation and quantification of seven polyols (erythritol, threitol, ribitol, arabinitol, xylitol, mannitol and galactitol) in serum. The technique used either packed (1.5 m long, 2.0 mm i.d., 3% SP-2340, 75% cyanopropyl-25% methyl silicone gum) or capillary (50 m long, 0.2 mm i.d., coated with OV-1701, crosslinked polar silicone gum) columns, with temperature programming (28). Starting with 0.5 ml serum, proteins were precipitated with methanol, the supernatant evaporated to dryness, and the peracetyl derivatives formed (with acetic anhydride-pyridine, 2:1, v/v). After stopping the reaction with water, the derivatives were extracted with hexane-chloroform, washed with water, dried with anhydrous Na_2SO_4, evaporated, and redissolved in hexane. Selected ion monitoring was used to detect and quantify the M-59 (loss of CH_3COO group) from the protonated molecules obtained in the positive CI (isobutane) mode: $m/z = 231$ (tetritols), $m/z = 303$ (pentitols), $m/z = 375$ (hexitols), and $m/z = 317$ (2-deoxygalactitol, int. std.). More than adequate chromatographic separation is available for the separation of individual tetritols, pentitols, and hexitols in the course of selected ion monitoring. Quantification was accomplished in essentially the same manner as in the original method.

Individual Pentitols. Another capillary GC/MS technique was developed for the separation and quantification of ribitol, xylitol, and total arabinitol *(29)*. Using, conveniently, the same sample preparation and derivatization (TMS), positive CI selected ion monitoring, and quantification techniques as in the original method, the separation of pentitols was accomplished isothermally at 195 °C using a 50 m long, 0.32 mm i.d. methyl silicone column (equivalent to OV-17). An important modification was incorporation of fully deuterated, D7-arabinitol as an internal standard (monitored $m/z = 520$).

Arabinitol Enantiomers. The first GC/MS technique for the separation of the enantiomers of arabinitol used custom-made 20-40 m long, 0.25 mm i.d. glass capillary columns coated with -perpentylated cyclodextrin *(30)*. The somewhat cumbersome glass capillary columns have been recently superseded by commercially available 30 m, 0.25 mm i.d. fused silica capillary columns, coated with permethylated β-cyclodextrin in a polysiloxane gum phase (β -Dex 120, Supelco, Inc. Bellefonte, PA), operated isothermally at 100 °C *(31)*. To protect the analytical column, it should be connected to the GC injector (on-column type, operated at room temperature) by a precolumn. Deactivated, uncoated, fused silica capillaries are efficient and inexpensive precolumns; they need to be changed after every 25-30 analyses but permit several hundred analyses before gradual loss of chromatographic resolution necessitates replacement of the analytical column. The exit end of the analytical column enters the mass spectrometer ion source through a transfer line maintained at 175 °C. Helium is used as the chromatographic carrier gas and isobutane, the CI reagent gas is admixed at the ion source.

Both positive *(27)* and negative CI mass spectrometry have been evaluated for the detection and quantification of the separated enantiomers. Negative CI, involving the selected ion monitoring of ions corresponding to the loss of an OCF_3CO group, at $m/z = 518$ for the enantiomers and $m/z = 525$ for the int. std., $(^2H_7)$D-arabinitol (96 atom % deuterium) offers significant advantages including increased sensitivity (major reduction in required sample size) and improved specificity *(32)*. Quantification of individual enantiomers, when needed, can be made as before using control (normal) samples supplemented with increasing quantities of D- and L-arabinitol. In most cases, however, D/L arabinitol ratios, rather than individual concentrations, are determined, using peak areas; the rationale for this is discussed later, together with results.

Filter Paper Sampling of Whole Blood and Urine. Collection of physiological fluids on filter paper is inherently simple *(31)*. The filter paper suggested is the kind used for phenylketonuria screening (Model #904, Schleicher & Schuell Inc., Keene, NH); based on limited experience, most other types of filter paper would also be adequate. There are two ways to place a whole blood sample on a filter paper. One might obtain a drop of blood using a lance to puncture a fingertip (or infant heel) and touch the filter paper until the ring (12 mm diameter) is filled. Alternatively, when venous blood is collected for some other purpose (e.g., in a red top tube), a drop of blood can be placed on the filter paper with a micropipet (20 uL). This method can also be used for blood serum and urine. Because only the D/L ratio is measured, the quantity of blood/urine placed on the filter paper need not be known. When the concentration of arabinitol, or some other polyol, is also to be determined, a calibrated micropipet is used to deposit a known amount (usually 20 uL) of fluid within the ring area on the paper. After collection, the filter papers are left to dry at room temperature (about 30 min); they can be stored at room temperature for at least two weeks, and may also be mailed in regular envelopes. Sample preparation has been kept simple: the sample spots, cut from the filter paper, are extracted with methanol, evaporated to dryness, and derivatized (37 °C, 45 min) using methylene chloride and trifluoroacetic anhydride. After evaporating the excessive reagent and the solvent to near dryness, residues are dissolved in toluene, and aliquots analyzed by GC/MS as described.

Results and Discussion: Methodology

Advantages of GC/MS. Straight GC techniques have several shortcomings *(33)*. Because identification is based on retention times, the ever present "chemical noise" decreases the reliability of positively identifying small peaks. In contrast, selected ion monitoring is much more specific because only ions characteristic of the analyte are monitored. Incremental sensitivity, i.e., an ability to determine small concentration differences is higher with selected ion monitoring than with flame ionization detectors. Using GC/MS makes it possible to use deuterium enriched D-arabinitol as the internal standard, thereby eliminating notorious interference problems of GC techniques which often necessitated the use of multiple internal standards *(24,25)*. Although the GC/MS techniques described may be carried out using electron ionization and monitoring appropriately selected ions of high abundance, chemical ionization with isobutane is more reliable and specific; no mass interference of any kind was encountered in more than a thousand analyses.

Enantiomer Separation and D/L Ratios. A novel combined microbiological and gas chromatographic technique for the quantification of the concentrations of arabinitol enantiomers in serum, urine, and tissue *(34)* provided important information about the stereoisomeric configuration of endogenous arabinitol; however it was too cumbersome for routine use. A subsequent GC method used a two-step procedure: first, total arabinitol was determined, next, the sample was treated with D-arabinitol dehydrogenase and the remaining L-arabinitol analyzed *(35)*. The next development was a multi-dimensional GC technique which applied a custom-made chiral column for the final separation of the enantiomers *(36)*. Although more practical than the previous approaches, and yielded some useful clinical data, this is still a unwieldy technique.

Bacterial D-arabinitol dehydrogenase is used to remove D-arabinitol (by oxidizing it to D-xylulose) in a spectrophotofluorimetric technique *(37,38)*. A problem with this technique is that endogenous mannitol, which is known to increase in renal dysfunction *(39)*, is also oxidized and thus included in the results *(40)*.

Figure 2 shows the separation and selected ion monitoring of the arabinitol enantiomers, and also that of the two other pentitols, using the negative CI GC/MS technique with direct chiral chromatography *(32)*. Complete separation of the four analytes was achieved in 7 minutes. Serum and urine were from the same cancer patient, who had no infection, showing D/L ratios of 2.8 for serum and 2.5 for urine; this profile is similar to those obtained for normal subjects. The concentrations of ribitol and xylitol are not determined routinely as they have no apparent diagnostic significance.

The most important analytical advantage of using D/L ratios instead of individual concentrations of the separated enantiomers is that there is no need for time-consuming calibration curves. Within-day reproducibility of ratio determinations is 5-10%. Although there is no need for the internal standard, deuterated arabinitol is still used because it's retention time, peak height, and peak shape are convenient controls of the overall analytical procedure. Another advantage of having a known amount of the internal standard present that it facilitates, if needed, a reasonably good (within 15%) quantification of individual enantiomers.

Filter Paper Sampling. The consequences of using filter paper for sampling was thoroughly tested *(31)*. Recovery of D-arabinitiol (2.8-92 ng range) from the filter paper with methanol extraction was 95-105% (n = 6). Reproducibility, storage, and transportation stability of D/L arabinitol ratios of serum, venous and fingerstick whole blood samples (n = 8) on filter paper were compared to serum standards analyzed directly. The results revealed that for all groups of filter paper samples, including those for storage (1, 3, 7, and 14 days) and transportation (roundtrip air from New York to Michigan): (a) standard deviations were equal or smaller than that of the serum standard; (b) means were statistically indistinguishable (t-test, independent samples) from that of the standard; (c) ratios of means of whole blood (venous and finger) over the mean of the serum standard; (d) the range of mean values was within ± 0.3 standard deviation of the mean of the standard. Comparable performance data were obtained for serum and urine samples from both normal subjects and patients with confirmed disseminated candidiasis.

The filter paper sampling technique has several obvious practical advantages: whole blood instead of serum can be analyzed, only one drop of whole blood, serum, or urine is needed, sample quantity need not be measured, filter papers can be stored at room temperature, and can be air transported.

Results and Discussion: Applications

Rationale for Using D/L Arabinitol Ratios. A serious specificity problem with the "arabinitol" technique, recognized from the beginning *(24-26)*, has been the fact that renal dysfunction, which is often concurrent with disseminated candidiasis, also produces increased serum arabinitol concentrations. It was suggested that using arabinitol/creatinine ratios instead of arabinitol concentrations would counter poor specificity in the presence of renal dysfunction *(25)*. Although arabinitol/creatinine ratios have been used in several GC applications *(41-45)*, the technique remained controversial; no correlation between arabinitol and creatinine was found in a study of patients on dialysis *(39)*.

The source and biochemical role, if any, of endogenous arabinitol in blood, urine, and cerebrospinal fluid is unkown. It has been suggested that arabinitol may be a metabolic product of arabinose, formed from the latter with the aid of NaDP and a dehydrogenase *(46-48)*. Based on an old case history about L-arabinitol in pentosuric urine *(49)*, it appeared convenient to assume that endogenous arabinitol would be of the L form. However, a combined microbiological and GC technique revealed that both enantiomers are present in human serum and urine, and that the ratio between the two enantiomers is approximately 1:1 *(34)*. Nevertheless, the fact that *Candida* species produce only D-arabinitol *(23, 24, 29)*, makes the prospect of using the enantiomers to circumvent the interference of renal dysfunction attractive. The development of chiral GC columns has made it possible to develop simple GC/MS technique for the separation and quantification of arabinitol enantiomers, and the diagnostic utility of using D/L ratios makes the technique even more convenient.

Endogenous Total Arabinitol and D/L ratios. Endogenous serum arabinitol concentration, i.e., total arabinitol, was determined using both GC and GC/MS by several investigators. Most values have been in the 0.2-0.5 ug/mL range; for example, in one GC/MS study on adult blood bank controls (n=25), endogenous arabinitol was 0.38 ± 0.12 ug/mL, range: 0.22-0.68 (29). Endogenous urinary arabinitol concentrations change significantly depending upon the volume of urine, and are best expressed with respect to creatinine because arabinitol is cleared by the kidney at virtually the same rate as creatinine; in one study, normal urinary excretion ratio was 67.2 ± 30.7 ug/mg creatinine (25).

Figure 3A shows serum and urine profiles with normal D/L arabinitol ratios. The figure also shows the monitoring of the internal standard for the serum sample. This is shown to illustrate the use of the internal standard which is added to every analysis. The results of the urine samples shown (and also the serum in Figure 2) include similar monitoring of the internal standard but these are omitted from the figures.

Data on DL arabinitol ratios in normal serum and urine are summarized in Table II. The upper limit of normal is defined as the mean+3 standard deviation. Accordingly, when the D/L is >3.3 in serum and/or >2.5 in urine, the attending physician should be alerted to the possibility of disseminated candidiasis, and new samples should be requested for confirmation. The agreement between D/L arabinitol ratios of serum and urine is remarkable, particularly in light of the fact that the concentration of total arabinitol is 50-60 times larger in normal urine than in normal serum. As illustrated in Figure 3B, urine from a patient with abnormal D/L ratio also shows abnormal D/L ratio. Work is in progress to evaluate the diagnostic value of analyzing both serum (whole blood) and urine, or urine alone, to further increase reliability.

It is noted that the D/L arabinitol ratio in the cerebrospinal fluid of "controls" was found to be approximately 10-fold higher than the ratio in normal serum and blood (33). The high ratio is attributed to D-arabinitol because the L-arabinitol concentration was shown to be nearly identical to that in serum. Similarly high ratios were found for a variety of diseases, but without *Candida* infection. While there is no immediate diagnostic use of this observation, some basic biochemistry work is indicated to explain this unexpected phenomenon.

Table II. Normal D/L Arabinitol in Serum and Urine

	n	Mean	St.Dev.	Range	Increased When
Serum	29	1.8	0.5	0.8-2.8	>3.3
Urine	14	1.6	0.3	1.1-2.1	>2.5

Upper limit of normal taken as mean + 3 St.Dev.
When ratio >upper limit: candidiasis suspected.

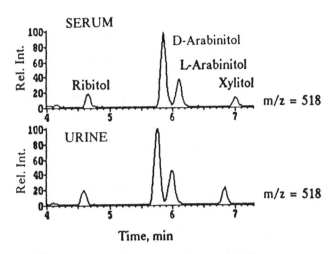

Figure 2. Chiral chromatographic separation and selected ion monitoring of the enantiomers of arabinitol and of the other two pentitols in serum and urine from a cancer patient without candidiasis.

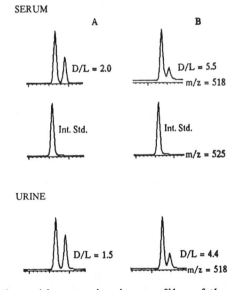

Figure 3. Selected ion monitoring profiles of the enantiomers of arabinitol in normal serum and urine (A) and in a patient with suspected disseminated candidiasis (B). The internal standard (deuterated arabinitol) is monitored in every analysis (not shown for urine).

Diagnostic Sensitivity and Specificity. Total pentitol concentrations, determined by the earlier GC/MS techniques, were applied to a study population of 1250 samples from 284 patients; results were summarized in reference *(32)*. That data indicated the desirability of developing a technique with greater specificity which led to the D/L ratio technique.

Diagnostic sensitivity is defined as the number of positives found/total number with the disease (positivity when infection is present). For the D/L ratio procedure, diagnostic sensitivity was 95% as determined in 22 patients having candidiasis confirmed by positive blood culture or autopsy (Table III). The one false negative gave normal D/L arabinitol ratio by both +CI and -CI. It is noted that only 68% of these cases had renal dysfunction.

Diagnostic specificity quantifies the performance of a test when the disease is not present. As discussed earlier, renal dysfunction seriously interferes when the diagnosis is based on the determination of arabinitol concentration. Indeed, previous arabinitol techniques, both GC and GC/MS, were nearly useless in the presence of renal dysfunction, yielding very poor diagnostic specificity. Accordingly, the specificity of the D/L arabinitol ratio technique was evaluated by selecting a population purposely biased, with 41% renal dysfunction, creatinine > 1.5 mg/dL *(32)*. The detection of six false positives yielded an 88% specificity Table III). Of the six false positives four were cancer patients, two with aspergillosis (an OI often associated with difficult to diagnose concurrent candidiasis), one with oral candidiasis, and one without any infection. The fifth patient was a renal transplant, the sixth a patient with serious renal dysfunction, both without known infection.

Table III. Diagnostic Sensitivity and Specificity

	Sensitivity	95%
Patients with confirmed candidiasis	n =	22
D/L arabinitol ratio >3.3 (range 3.3->50)		21
False negative: D/L=1.2		1
	Specificity	88%
Patients without candidiasis	n =	49
Purposely biased population:		
41% with renal dysfunction		
D/L arabinitol ratio normal, <3.3)		43
False positives, D/L>3.3		6

Monitoring D/L Arabinitol Ratios. Two potentially useful clinical applications are the serial monitoring of patients at risk and of patients undergoing antifungal chemotherapy. Figure 4 shows the monitoring of a cancer patient at risk. In this case, the patient never developed candidiasis. Figure 5 shows the monitoring of a patient at risk who did develop disseminated candidiasis, as confirmed by autopsy. However, blood cultures remained negative during the period of increased D/L arabinitol values. The decrease of D/L ratios have also been demonstrated in the course of successful antifungal chemotherapy, although the number of cases to-date is

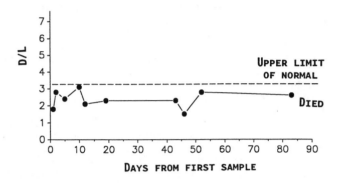

Figure 4. Serial monitoring of serum D/L arabinitol in a cancer patient who was at high risk of developing disseminated candidiasis but never did.

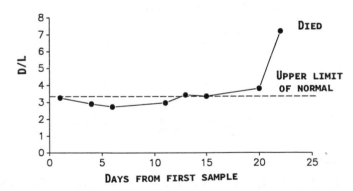

Figure 5. Serial monitoring of serum D/L arabinitol in a cancer patient who did develop disseminated candidiasis which was confirmed upon autopsy.

too small for statistical evaluation. The advantages of monitoring are obvious: early diagnosis will permit treatment while the fungus load is still small, and monitoring chemotherapy will assist the evaluation of the regression/progression of the disease during treatment.

Summary

Analytical Methodology: (a) In addition to serum and urine, whole blood may also be used directly; (b) Sample drops of unmeasured size may be placed as spots on filter paper which can be kept for at least two weeks at room temperature and may be mailed; (c) The D and L enantiomers of arabinitol are separated on chiral columns and detected without matrix interference (high analytical specificity) using chemical ionization and selected ion monitoring; (d) Only D/L arabinitol ratios are determined, thus no calibration is needed; (e) Both whole blood/serum and urine may be analyzed from the same patient for increased reliability; (f) Ten samples/day can be analyzed routinely without special requirements; single samples can be analyzed in two hours.

Clinical Performance: (a) Endogenous D/L arabinitol ratios, and upper limits of normal values have been determined in serum and urine; (b) Diagnostic sensitivity was 95% (one false negative in 22 cases); (c) To test the ability of the method to distinguish between increased arabinitol due to candidiasis and renal dysfunction, diagnostic specificity was determined in an uninfected population heavily biased (50%) to patients with renal dysfunction; specificity was 88% (six false positives in 49 cases); (d) In an ongoing pilot study, serial monitoring has been commenced and initial data obtained for small groups of patients at high risk of developing disseminated candidiasis, and for patients undergoing therapeutic or prophylactic antifungal chemotherapy.

Current new applications include the semi-routine monitoring of bone marrow transplantation patients, providing diagnostic support for newborns with suspected infection, confirmatory analyses and monitoring of antifungal chemotherapy of pediatric oncology patients, and the initiation of double-blind inter-institutional comparative studies utilizing the filter paper technique.

Acknowledgments. This work was supported by the T.J. Martell Memorial Foundation for Leukemia, Cancer and AIDS research.

Literature Cited

1. Horn, R.; Wong, B.; Kiehn, T.; Armstrong, D. *Rev.Infect.Dis.* 1985, 7, 646.
2. Meunier, F.; Aoun, M.; Bitar, N. *Clin.Infect.Dis.* 1992, 14, S120.
3. Maksymiuk, A.; Thorngprasert, S.; Hopfer, R.; Luna, M.; Fainstein, V.; Bodey, G. *Am.J.Med.* 1984, 77(4d), 20.
4. Hughes, W. *Pediatr.Infect.Dis.J.* 1982, 1, 11.
5. Jarvis, W. *3rd International Conference for Noscomial Infections* 1990.
6. Leibovitz, E.; Rigaud, M.; Sulachni, C.; Kaul, A.; Greco, M.; Pollack, H.; Lawrence, R.; Di John, D.; Hanna, B.; Krasinski, K.; Borkowsky, W. *Pediatr.Infect.Dis.J.* 1991, 10, 888.
7. Sobel, J. Controversial Aspects of Candidiasis in the Acquired Immuno-deficiency Syndrome. In: *Mycoses in AIDS Patients*, Bossche, H.; Mackenzie,D.; Cauwenbergh,G.; Van Cutsem,J.; Drouhet,E. and Dupont,B. Ed.; New York: Plenum Press, 1990, p. 93-100.

8. Hughes, W.; Armstrong, D.; Bodey, G.; Feld, R.; Mandell, G.; Meyers, J.; Pizzo, P.; Schimpff, S.; Shenop, J.; Wade, J.; Young, L.; Wow, M. *J.Inf.Dis.* **1990**, *161*, 381.
9. Francis, P.; Walsh, T. *Oncology* **1992**, *6*, 81.
10. Francis, P.; Walsh, T. *Oncology* **1992**, *6*, 133.
11. Gooman, J.; Winston, D.; Greenfiled, R.; Chandrasekar, P.; Fox, B.; Kaizer, H.; Shadduck, R.; Shea, T.; Stiff, P.; Friedman, D.; Powderly, W.; Silber, J.; Horowitz, S.; Weisdorf, D.; Ho, W.; Gilbert, G.; Buell, D. *New Eng.J.Med.* **1992**, *326*, 845.
12. Odds, F. *Candida and Candidosis*; Ed. 2nd , Bailliere Tindall: London, 1988;
13. Dupont, B. Clinical Manifestations and Management of Candidosis in the Compromised Patient. In: *Fungal Infection in the Compromised Patient*, Warnock, D.;W.;; Richardson, M.;D.; Eds.: by New York: John Wiley and Sons, 1991, p. 56-83.
14. Rippon, J. *Medical Mycology*; Ed. 3rd, Saunders: Philadelphia, 1988, p.532.
15. Shepherd, M.; Poulter, R.; Sullivan, P. *Annu.Rev.Microbiol.* **1985**, *39*, 579.
16. Meunier, F. *Cancer Invest.* **1991**, *9*, 151.
17. Bennett, J.E. *Rev.Inf.Diseases* **1987**, *9*, 398.
18. Burnie, J. ; nd Matthews, R. Serodiagnosis of Systematic Mycoses. In: *Handbook of Serodiagnosis in Infectious Diseases*, Matthews, R.;; Burnie, J.; Eds.: by Butterworth and Heinemann, 1991, p. 205-216.
19. de Repentigny, L. *Clin.Infect.Dis.* **1992**, *14*, S11.
20. Ness, M.; Waughan, W.; Woods, G. *J.Inf.Dis.* **1989**, *159*, 495.
21. Fujita, S.; Hashimoto, T. *J.Clin.Microbiol.* **1992**, *30*, 3132.
22. Walsh, T.; Hathorn, J.; Sobel, J.; Merz, W.; Sanchez, V.; Maret, M.; Buckley, H.; Pfaller, M.; Schaufele, R.; Sliva, C.; Navarro, E.; Lecciones, J.; Chandrasekar, P.; Lee, J.; Pizzo, P. *New Eng.J.Med.* **1992**, *324*, 1026.
23. Bernard, E.; Christiansen, K.; Tsang, S.; Kiehn, T.; Armstrong, D. *J.Clin.Microbiol.* **1981**, *14*, 189.
24. Kiehn, T.; Bernard, E.; Gold, J.; Armstrong, D. *Science* **1979**, *206*, 577.
25. Wong, B.; Bernard, E.; Gold, J.; Fong, D.; Armstrong, D. *J.Inf.Dis.* **1982**, *146*, 353.
26. Roboz, J.; Suzuki, R.; Holland, J.F. *J.Clin.Microbiol.* **1980**, *12*, 594.
27. Roboz, J.; Nieves, E.; Holland, J.F. *J.Chrom.* **1990**, *500*, 413.
28. Roboz, J.; Kappatos, D.; Greaves, J.; Holland, J.F. *Clin.Chem.* **1984**, *30*, 1611.
29. Roboz, J.; Kappatos, D.; Holland, J.F. *Europ.J.Clin.Microbiol.* **1987**, *6*, 708.
30. Konig, W.; Mischnick-Lubbecke, S.; Brassat, B.; Lutz, S.; Wenz, g. *Carbohydr.Res.* **1988**, *183*, 11.
31. Roboz, J.; Yu, Q.; Holland, J.F. *J.Microbiol.Methods* **1992**, *15*, 207.
32. Roboz, J.; Katz, R. *J.Chrom.* **1992**, *575*, 281.
33. Roboz, J. Gas chromatographic and mass spectrometric techniques for the diagnosis of disseminated candidiasis. In: *Methods in Analytical Microbiology: Chromatography and Mass Spectrometry*, Fox, A.; Larsson, L.; Morgan, S.; and Odham, G. Eds.; New York: Plenum Press , 1990, p. 239-258.
34. Bernard, E.; Wong, B.; Armstrong, D. *J.Inf.Dis.* **1985**, *151*, 711.
35. Wong, B.; Brauer, K. *J.Clin.Microbiol.* **1988**, *26*, 1670.
36. Wong, B.; Castellanos, M. *J.Chrom.* **1989**, *495*, 21.
37. Soyama, K.; Ono, E. *Clin.Chim.Acta* **1985**, *149*, 149.
38. Soyama, K.; Ono, E. *Clin.Chim.Acta* **1987**, *168*, 259.
39. Roboz, J.; Kappatos, D.; Holland, J.F. *Clin.Chem.* **1990**, *36*, 2082.
40. Soyama, K.; Ono, E. *Clin.Chem.* **1988**, *34*, 432.

41. Eng, R.; Chmel, H.; Buse, M. *J.Inf.Dis.* **1981**, *143*, 677.
42. Deacon A.G, *J.Clin.Pathol.* **1986**, *39*, 842.
43. Wells, C.; Sirany, M.; Blazevic, D. *J.Clin.Microbiol.* **1983**, *18*, 353.
44. Wong, B.; Baughman, R.; Brauer, K. *J.Clin.Microbiol.* **1989**, *27*, 1859.
45. Wong, B.; Brauer, K.; Clemens, J.; Beggs, S. *Infect.Immunity* **1990**, *58*, 283.
46. Veiga, L.A.; *Biochem.Biophys.Res.Comm.* **1960**, *2*, 440.
47. Veiga, L.A.; *J.Gen.Appl.Microbiol.* **1968**, *14*, 65.
48. Barnett, J.; *Adv.Carbohydr.Res.* **1976**, *32*, 125.
49. Touster, O.; *J.Biol.Chem.* **1958**, *230*, 1031.
50. Christensson, B.; Roboz, J. *J.Neurol.Sciences* **1991**, *105*, 234.

RECEIVED June 15, 1993

Chapter 10

Ribonucleic Acid Modification in Microorganisms

Charles G. Edmonds[1,3], Ramesh Gupta[2], James A. McCloskey[1], and Pamela F. Crain[1]

[1]Department of Medicinal Chemistry, University of Utah, Salt Lake City, UT 84112
[2]Department of Medical Biochemistry, Southern Illinois University School of Medicine, Carbondale, IL 62901

Ribonucleic acid contains more than 80 modified nucleosides that play multiple functional roles, including regulation of gene expression, adaptation to stress and promotion of fidelity of protein synthesis. Transfer RNA is the most highly modified RNA, and serves as a rich source of modified nucleosides that may be restricted to one or another of the archaeal (formerly archaebacterial (1)), bacterial, or eucaryal phylogenetic domains. The presence (or absence) of certain modified nucleosides can thus serve as phylogenetic markers. Nuclease digests of tRNA (about one nanomole) can be analyzed directly as mixtures using thermospray liquid chromatography/mass spectrometry (LC/MS) to characterize modified nucleosides from both their mass spectra and chromatographic retention times.

Mass spectrometry has been used extensively for structure characterization of new nucleosides in nucleic acids (2,3). Directly-combined high performance liquid chromatography/mass spectrometry (LC/MS) (4,5) has proven an especially powerful screening technique leading to the identification of trace amounts of modified nucleosides, and in favorable cases permitted their complete structural characterization without the necessity of isolation of individual components (6-9).

All classes of nucleic acid, e.g. DNA and ribosomal (rRNA), messenger (mRNA), and transfer (tRNA) RNAs contain posttranscriptional modifications. DNA (except for that of bacteriophages) contains mainly three modified base constituents, N^4- or 5-methylcytosine and N^6-methyladenine, and their occurrence is not phylogenetically restricted. Modified nucleosides are present in the small (16S–18S) and large (23S–28S) rRNAs (reviewed in ref. 10), but archaeal rRNA is insufficiently well characterized to permit an assessment of its utility for phylogenetic delineation.

[3]Current address: Battelle, Pacific Northwest Laboratories, Richland, WA 99352

0097–6156/94/0541–0147$06.00/0
© 1994 American Chemical Society

Although mRNA, at least in eucarya, contains low levels of nucleoside modification, it is difficult to isolate and is present in amounts too small to be analytically useful. In contrast, transfer RNA (tRNA) is readily isolated, and is extensively modified, and is thus the most informative for studies of phylogenetic relationships and patterns of modification.

Structure of Transfer Ribonucleic Acid (tRNA)

tRNA is the smallest of the cellular RNAs, having a length of between 75 and 90 nucleotides. The molecule comprises four major nucleosides (Figure 1), and in addition, a large number of modified nucleosides produced by posttranscriptional enzymatic reactions. Some of these modifications can be as simple as methyl substitution, while others involve extensive remodeling of the basic purine ring, e.g. queuosine (7) and archaeosine (8). The pyrimidine and purine ring numbering systems are shown for cytidine and adenosine, respectively, to permit visualization of nucleoside structures in discussions to follow. The names, abbreviations and structures of modified nucleosides which have been identified in sequenced tRNAs have been compiled (11).

Occurrence of Modified Nucleosides in tRNA. More than 80 modified nucleosides have been characterized in tRNA from organisms in all three primary phylogenetic domains. Both the sequence location and chemical identity of certain modified nucleosides can be highly restricted. Examination of the "cloverleaf" representation of a composite tRNA molecule shown in Figure 2 reveals that the modified nucleosides are concentrated in specific regions and sequence locations in the molecule. The most extensively modified location in the molecule is position 34, the first nucleoside in the anticodon region (which interacts with the third codon position of the messenger RNA). An unusually large number of modified nucleosides are also found at position 37, adjacent to the last anticodon nucleoside.

Roles of Modified Nucleosides in tRNA. Major roles for nucleoside modification in tRNA are the maintenance of translational fidelity and efficiency of codon usage during protein biosynthesis (reviewed in refs. 12 and 13). The nucleosides in positions 34 and 37 are particularly important in this regard. Another function of nucleoside modification may be stabilization of the tertiary structure as an adaptation to cellular stress, particularly during growth of thermophilic organisms at high temperatures. For example, the overall extent of tRNA methylation in *Bacillus stearothermophilus* increases with increasing growth temperature (14), and a survey (by LC/MS) of modified nucleosides in tRNA from a number of hyperthermophilic archaea showed a strong correlation between the content of ribose-methylated nucleosides and the temperature of optimal growth (8), evidently reflecting increased conformational stability provided by ribose methylation (15).

Phylogenetic Significance of Modified Nucleosides in tRNA

Living organisms are presently considered to belong to one of three primary phylogenetic domains: the archaea (formerly archaebacteria), bacteria (formerly eubacteria)

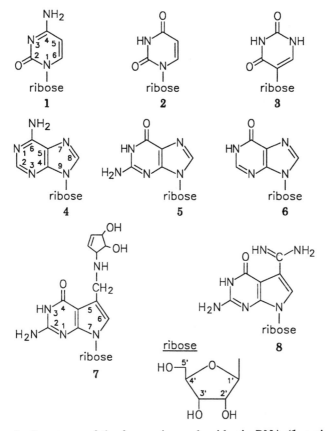

Figure 1. Structures of the four major nucleosides in RNA (**1**, cytidine (C); **2**, uridine (U); **4**, adenosine (A); **5**, guanosine (G)), and three common modified nucleosides (pseudouridine (**3**; Ψ), inosine (**6**; I) and queuosine (**7**; Q)). Ring numbering systems are illustrated for pyrimidines (**1**) and purines (**4**).

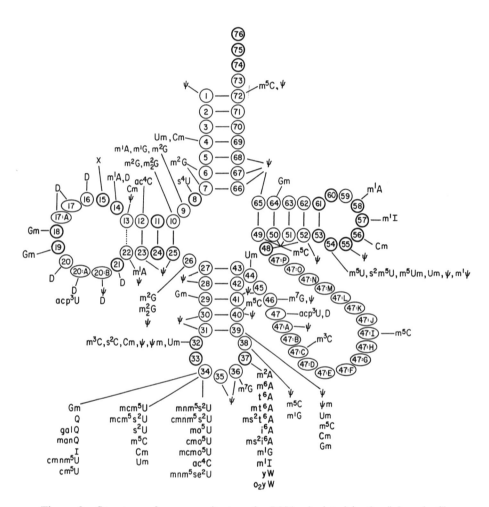

Figure 2. Structure of a composite transfer RNA, depicted in the "cloverleaf" form dictated by Watson-Crick base pairing. Nucleosides are numbered beginning from the 5'-end. (Reproduced from ref. 3. Copyright 1991 American Chemical Society.)

or the eucarya (*1*). The interrelationships among the domains are commonly represented in a tree-like configuration (Figure 3) with a root between the bacteria and archaea. These phylogenetic assignments are based on 16S rRNA sequences (*16*).

Comparative analyses of the occurrence of modified nucleosides in organisms in all three primary domains revealed that certain nucleosides are characteristically restricted to one or two domains (*17–19*). In some cases, sequence location of a common modified nucleoside may be highly conserved. For example, 2′-*O*-methylcytidine is found in all archaeal tRNA sequences at position 56, which is otherwise never modified in the other two domains (*11*).

Occurrence of Modified Nucleosides in the Primary Phylogenetic Domains. Among reasonably abundant (e.g. 0.1–1% occurrence) and widely-distributed modified nucleosides, 7-formamidino-7-deazaguanosine (archaeosine, **8**) (*20*) and m$^1\Psi$ (*21*) are two relatively abundant nucleosides that are restricted to archaeal tRNA. It should be noted, however, that not every archaeon contains **8**. 2-Methyladenosine and 3-methylcytidine, and N^6-isopentenyladenosine, are restricted to bacterial and eucaryal tRNA, respectively. Queuosine is widely distributed among both bacteria and eucarya, but is notably absent in yeast.

Modified Nucleosides Useful as Markers of Phylogenetic Domain. The composite tRNA structure (Figure 2) shows that a large number of modified nucleosides are present in the first anticodon position (nucleoside 34). A given nucleoside, however, is often restricted to a specific tRNA among perhaps 70 or so tRNA species each present in different amounts in the cell. Such a nucleoside might be phylogenetically restricted but difficult to analyze due to its low level of occurrence in the mixed tRNA pool. Several additional modified nucleosides, present in relatively low amounts and in unknown sequence locations, are also present in tRNA from hyperthermophilic archaea (*8*).

More generally useful are relatively abundant nucleosides which are easily identified in mixed tRNA. Table I compiles a number of these nucleosides from earlier reviews (*17–19*). Chromatographic retention times are catalogued elsewhere (*5*). In addition to the presence of the uniquely phylogenetically-associated nucleosides discussed above, the absence of a certain modified nucleoside can also provide useful information. An assay for several nucleosides will prove generally reliable, but the conclusions obviously depend on the quality of data from which they are derived, particularly with regard to a negative finding.

Determination of Modified Nucleosides in tRNA by Directly-Combined Liquid Chromatography/Mass Spectrometry

During early studies to establish the sequence of pure (isoaccepting) tRNA species, the presence of modified nucleosides was commonly determined from 2-D thin-layer chromatography, more recently with improved sensitivity from ^{32}P-post-labelling techniques (*23*). Identification is based solely on chromatographic mobility, and requires skill in recognizing locations of the modified substituents. High performance liquid chromatography with detection by UV absorbance provides an additional dimension of identification of modified nucleosides that, in addition to

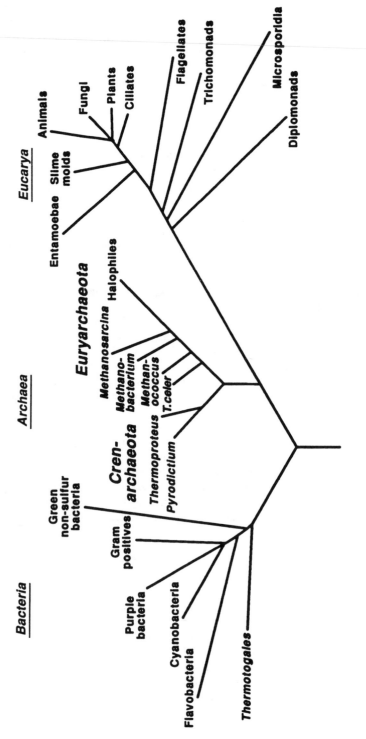

Figure 3. Rooted phylogenetic tree for the three primary domains of living organisms. (Adapted with permission from ref. 16. Copyright 1993 Federation of American Societies for Experimental Biology.)

chromatographic mobility, can provide either the ratio of absorbance at two wavelengths (e.g. 254 and 280 nm) (*24*), or full UV spectra (*25*). However, directly-combined HPLC-mass spectrometry (LC/MS) (*4,5,26*) provides in addition to elution times and UV absorbance characteristics, molecular weight and structural information, and is thus a more powerful experimental approach than either technique used alone. Essential features of the analysis are isolation and digestion of the tRNA to nucleosides (the smallest structural unit in which modifications are expressed), chromatographic separation of the components of the digested RNA, and on-line acquisition of mass spectra of the fractionated nucleosides.

Preparation of tRNA Digests. tRNA can be readily be isolated from cell extracts by a variety of simple protocols (*8,24,27*), and is commonly recovered from a 2M (NaCl or LiCl) salt-soluble fraction by precipitation with ethanol. Additional sample treatment, and use of nucleases P1 and phosphodiesterase I, and alkaline phosphatase, to digest RNA to nucleosides in a form suitable for mass spectrometric analysis have been described (*28*). The nucleoside mixture can be analyzed directly without removal of the enzymes. A 25–50 µg sample (1–2 nanomoles) is typically suitable for analysis.

Reversed Phase HPLC. Volatile buffers are a prerequisite for mass spectrometry of HPLC eluents. The ammonium acetate/acetonitrile-water buffer system most extensively used for analysis of tRNA digests (*5*) was adapted from a published separation devised for a gradient controller permitting exponential solvent programs (*24*). Separation is achieved on standard 250 x 4.6 mm C-18 columns, at a flow rate of 2 mL/min; an extensive listing of elution times of modified nucleosides using this solvent system has been published (*5*). It should be noted that if other LC/MS interfaces (*26*) that require lower flow rates are used, the elution times will change unless a column with a narrower i.d. (with a proportional reduction in flow rate) is used. The relative order of elution, however, should be unaffected.

Chromatograms from reversed phase HPLC of unfractionated tRNA from *Salmonella typhimurium* and several purified (isoaccepting) *Escherichia coli* tRNAs (both bacteria) can be found in ref. 24; the eluents utilized are volatile, and compatible with LC/MS. Examples of reversed phase separations of unfractionated tRNA from brewer's yeast and bovine liver (both eucarya), as well as *E. coli,* in a phosphate buffer-based eluant have been published (*25*). Elution order in the latter system (*25*) is generally the same as the former (*24*) with the exception of positively-charged nucleosides (e.g. m^{7}G; m^{1}A) which elute relatively earlier than with the ammonium acetate buffer system.

The chromatogram from an LC/MS analysis of digested tRNA from the hyperthermophilic archaeon, *T. neutrophilus,* is shown in Figure 4 and illustrates the complexity of the nucleoside mixture. Mass spectra from the unnumbered peaks did not yield ions characteristic of nucleosides, and they probably represent contaminants from isolation procedures and enzyme preparations. Deoxynucleosides arise from trace amounts of DNA contaminants co-extracted with the tRNA; their presence does not interfere with mass spectrometric analysis of the RNA nucleosides because they can readily be recognized from their thermospray mass spectra (see below).

Table I. Selected Phylogenetically Significant Modified Nucleosides
in Sequenced tRNAs From the Three Primary Domains

Nucleoside (MH^+, BH_2^+ ions)	Archaea	Bacteria	Eukarya
2-methyladenosine (281,150)	–	+	–
N^6-isopentenyladenosine (336,204)	–	–	+
3-methylcytidine (258,126)	–	–	+
5-methylcytidine (258,126)	+	–	+
2-thiocytidine (260,128)	+	+	–
N^2-methylguanosine (298,166)	+	–	+
$N^2,N^2,O\text{-}2'$-trimethylguanosine (312,180)	+	–	–
archaeosine, **8** (325,193)	+	–	–
1-methylpseudouridine (259, [a])	+	–	–
queuosine, **7** (410; [a,b])	–	+	+
3-(3-amino-3-carboxypropyl)uridine (346,214)	–	+	+
4-thiouridine (261,129)	+	+	–

[a] BH_2^+ generally not observed
[b] Additional characteristic ions in ref. 22

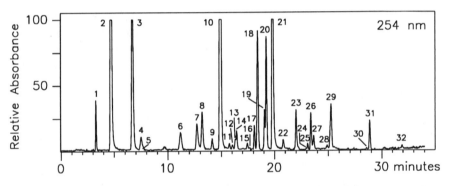

Figure 4. Chromatogram (detection by 254 nm UV absorbance) from reversed-phase chromatographic separation of nucleosides from LC/MS analysis of an enzymatic digest of 15 μg of unfractionated tRNA from the hyperthermophilic archaeon, *Thermoproteus neutrophilus*. Structures and chemical names are given in ref. 11. Peak identities are: 1, structure **3**; 2, **1**; 3, **2**; 4, m¹Ψ; 5, dC; 6, m⁵C; 7, m¹A; 8, Cm; 9, **6**; 10, **5**; 11, m⁷G; 12, s⁴U; 13, Um; 14, dG; 15, dT; 16, m⁵Cm; 17, m¹I; 18, m¹G and Gm; 19, ac⁴C; 20, m²G; 21, **4**; 22, dA; 23, m₂²G; 24, Am; 25, t⁶A; 26, **8**; 27, ac⁴Cm; 28, s²Um; 29, m⁶A; 30, ms²t⁶A; 31, m₂²Gm; 32, mimG. (Reproduced with permission from ref. 8. Copyright 1991 American Society for Microbiology.)

Thermospray Mass Spectra of Nucleosides. Thermospray ionization of nucleosides produces two main ion types, the protonated molecular ion (MH^+) and an ion corresponding to the protonated free base (BH_2^+) produced via transfer of a proton from ribose to the protonated base and cleavage of the glycosidic bond (*4,5*). The latter ion type is commonly present as the most abundant ion in the mass spectrum. These ion types are characteristic of ions produced by other "soft ionization" methods, e.g. chemical ionization (*29*) and fast atom bombardment (*30*) and may be expected in mass spectra acquired using other LC/MS techniques (reviewed in ref. 26). An ion, $(S - H)^+$, corresponding to the ribose moiety (S) minus a proton, or more commonly $(S - H) \cdot NH_4^+$ is occasionally produced.

Thermospray mass spectra are shown in Figure 5 for two representative modified tRNA nucleosides, the methylated guanosine isomers, N^2-methylguanosine (m^2G) and 2'-*O*-methylguanosine (Gm). These spectra exhibit the typical ion types discussed above. The molecular weights of both nucleosides are the same, so both produce an MH^+ ion at *m/z* 298. In the spectrum shown in the upper panel, a prominent ion of *m/z* 166 is evident; there is a 132 u difference in mass between this ion and the MH^+ ion, corresponding to loss of the normal ribose moiety. The *m/z* 166 ion is assigned as the BH_2^+ ion, and the methyl group is therefore located in the base. The spectrum in the lower panel of Figure 5 shows prominent ions of *m/z* 298 and 152; the difference in mass between them, 146 u, corresponds to a methylribose species, allowing assignment of the structure of this component as Gm. The exact location of the methyl group is not determined from the mass spectrum, rather it is placed on the 2'-hydroxyl because the 3'- and 5'- oxygens of ribose participate in the phosphodiester linkage of the RNA backbone and are unavailable for methyl substitution and because it is presently the only known site of ribose methylation in RNA.

Modified nucleosides with simple substituents (methyl; substitution of S for O) generally show the major ion types described above. On the other hand, thermospray mass spectra of so-called "hypermodified" nucleosides such as **7** and t^6A (*N*-[[(9-β-D-ribofuranosyl-9*H*-purin-6-yl)amino]carbonyl]threonine (*5*), which have complex side chains, show extensive fragmentation involving the substituent, and so the MH^+ ion may not be observed. On the other hand, pseudouridine nucleosides, which have C–C glycosidic bonds, do not form the BH_2^+ ion type owing to the greater strength of this bond relative to the N–C bond in normal nucleosides. This ion type is also of very low abundance in **7** and **8**, which contain the 7-deazaguanine base moiety which also leads to increased glycosidic bond strength. Some examples of thermospray mass spectra of modified nucleosides are published in refs. 4, 5, 8 and 22.

Directly-Combined LC/MS. The preponderance of applications of LC/MS to analysis of mixtures of nucleosides from tRNA have utilized the thermospray interface 4,5,7–9), although the basic approach is adaptable to other LC/MS interfaces (*26*). The instrument configuration (*4,5*) consists of a liquid chromatograph coupled directly to the mass spectrometer. A dual wavelength UV monitor is placed in series between the two instruments and its output is digitized by the computer which controls mass spectrometric data acquisition. Mass spectra can be acquired continually throughout the separation, typically one scan every 1–2 seconds over the mass range 100 to 360. Usable mass spectra can routinely be obtained from nucleosides present at the tens of nanograms amount. If selected ion monitoring is

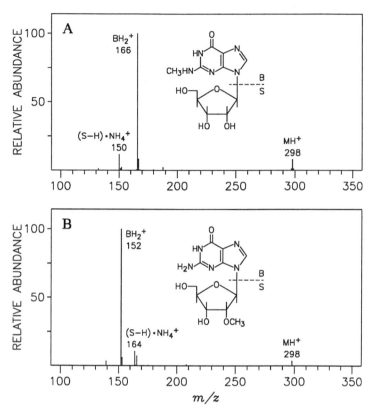

Figure 5. Thermospray mass spectra of (A) N^2-methylguanosine, and (B) 2′-O-methylguanosine, acquired from LC/MS analysis of a digest of tRNA from *Haloferax volcanii*.

performed, in which one or two characteristic ions are recorded for each component, sensitivity of detection at sub-nanogram amounts can be achieved (4).

The capabilities of LC/MS for analysis of nucleoside mixtures is exemplified by the analysis shown in Figure 6. The left panel shows a portion of the chromatogram from separation of a digest of mixed tRNA from *Haloferax volcanii* (formerly named *Halobacterium volcanii*). The right panel shows reconstructed ion chromatograms for ions characteristic of three methylated guanosine isomers which elute in this region. BH_2^+ ions characteristic of the unmodified guanine base moiety, monomethylguanine and dimethylguanine are m/z 152, 166 and 180, respectively. The MH^+ ions for monomethylguanosine and dimethylguanosine are m/z 298 and 312, respectively. The time vs. abundance profiles of these five ions show the added specificity gained from mass spectral analysis. The methylguanosine isomers m^1G and Gm are not visibly (e.g. using UV detection) separated during reversed-phase HPLC (5,24). The time vs. abundance profiles for the 152 and 166 ions do not precisely overlap, however, and so reveal the slight degree of separation between m^1G and Gm. (The thermospray mass spectrum of Gm, shown in the lower panel of Figure 5, in fact contains a weak 166 ion arising from incomplete separation from m^1G, which closely precedes it).

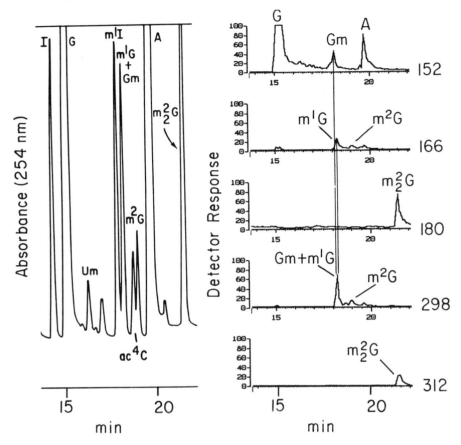

Figure 6. LC/MS analysis of tRNA from *Haloferax volcanii*. Left panel: partial chromatogram (UV detection at 254 nm). Right panel: time vs. abundance profiles (ion chromatograms) reconstructed from full-scan mass spectra. (Adapted with permission from ref. 4. Copyright 1985 Oxford University Press.)

Acknowledgment. Work from the authors' laboratory was supported by grants from the National Institutes of Health to J.A.M. (GM 21584 and GM 29812) and R.G. (GM 36042).

Literature Cited

1. Woese, C. R.; Kandler, O.; Wheelis, M. L. *Proc. Natl. Acad. Sci. USA* **1990**, 87, 4576–4579.
2. Crain, P. F.; Hashizume, T.; Nelson, C. C.; Pomerantz, S. C.; McCloskey, J. A. In *Biological Mass Spectrometry;* Burlingame, A. L.; McCloskey, J. A., Eds.; Elsevier Science Publishers: Amsterdam, 1989, 509–525.
3. McCloskey, J. A. *Accounts Chem. Res.* **1991**, 24, 81–88.
4. Edmonds, C. G.; Vestal, M. L.; McCloskey, J. A. *Nucleic Acids Res.* **1985**, 13, 8197–8206.
5. Pomerantz, S. C.; McCloskey, J. A. *Methods Enzymol.* **1990**, 193, 796–824.

6.   Edmonds, C. G.; Crain, P. F.; Hashizume, T.; Gupta, R.; Stetter, K. O.; McCloskey, J. A. *J. Chem. Soc., Chem. Commun.* **1987,** 909–910.
7.   McCloskey, J. A.; Edmonds, C. G.; Gupta, R.; Hashizume, T.; Hocart, C. H.; Stetter, K. O. *Nucleic Acids Res. Symp. Series* **1988,** *20,* 45–46.
8.   Edmonds, C. G.; Crain, P. F.; Gupta, R.; Hashizume, T.; Hocart, C. H.; Kowalak, J. A.; Pomerantz, S. C.; Stetter, K. O.; McCloskey, J. A. *J. Bacteriol.* **1991,** *173,* 3138–3148.
9.   Bangs, J. D.; Crain, P. F.; Hashizume, T.; McCloskey, J. A.; Boothroyd, J. C. *J. Biol. Chem.* **1992,** *267,* 9805–9815.
10. Björk, G. In *Processing of RNA;* Apirion, D., Ed.; CRC Press, Inc.: Boca Raton, FL, 1984, pp 291–330.
11. Sprinzl, M.; Hartmann, T.; Weber, J.; Blank, J.; Zeidler, R. *Nucleic Acids Res.* **1989,** *17* (Suppl.), r1–r172.
12. Dirheimer, G. In *Modified Nucleosides and Cancer;* Nass, G., Ed.; Springer-Verlag: Berlin Heidelberg, 1983, 15–46.
13. Björk, G. In *Transfer RNA in Protein Synthesis;* Hatfield, D. L.; Lee, B. J.; Pirtle, R. M., Eds.; CRC Press, Inc., Boca Raton, Fl, 1992, 23–85.
14. Agris, P. F.; Koh, H.; Söll, D. *Arch. Biochem. Biophys.* **1973,** *154,* 277–282.
15. Kawai, G.; Yamamoto, Y.; Kamimura, T.; Masegi, T.; Sekine, M.; Hata, T.; Iimori, T.; Watanabe, T.; Miyazawa, T.; Yokoyama, S. *Biochemistry* **1992,** *31,* 1040–1046.
16. Woese, C. R.; Olsen, G. J. *FASEB J.* **1993,** *7,* 113–123.
17. Gupta, R.; Woese, C. R. *Current Microbiol.* **1980,** *4,* 245–249.
18. Björk, G. *Chem. Scripta* **1985,** *26B,* 91–95.
19. McCloskey, J. A. *System. Appl. Microbiol.* **1985,** *7,* 246–252.
20. Gregson, J. M.; Crain, P. F.; Edmonds, C. G.; Gupta, R.; Hashizume, T.; Phillipson, D. W.; McCloskey, J. A. *J. Biol. Chem.,* **1993,** in press.
21. Pang, H.; Ihara, M.; Kuchino, Y.; Nishimura, S.; Gupta, R.; Woese, C. R.; McCloskey, J. A. *J. Biol. Chem.* **1982,** *257,* 3589–3592.
22. Phillipson, D. W.; Edmonds, C. G.; Crain, P. F.; Smith, D. L.; Davis, D. R.; McCloskey, J. A. *J. Biol. Chem.* **1987,** *262,* 3462–3471.
23. Kuchino, Y.; Hanyu, N.; Nishimura, S. *Methods Enzymol.* **1987,** *155,* 379–396.
24. Buck, M.; Connick, M.; Ames, B. N. *Anal. Biochem.* **1983,** *129,* 1–13.
25. Gehrke, C. W.; Kuo, K. C. *J. Chromatogr.* **1989,** *471,* 3–36.
26. *Liquid Chromatography/Mass Spectrometry: Techniques and Applications;* Yergey, A. L.; Edmonds, C. G.; Lewis, I. A. S.; Vestal, M. L., Eds.; Plenum Press: New York, NY, 1990, pp 89–125.
27. Reddy, D. M.; Crain, P. F.; Edmonds, C. G.; Gupta, R.; Hashizume, T.; Stetter, K. O.; Widdel, F.; McCloskey, J. A. *Nucleic Acids Res.* **1992,** *20,* 5607–5615.
28. Crain, P. F. *Methods Enzymol.* **1990,** *193,* 782–790.
29. Wilson, M. S.; McCloskey, J. A. *J. Am. Chem. Soc.,* **1975,** *97,* 3436–3444.
30. Crow, F. W.; Tomer, K. B.; Gross, M. L.; McCloskey, J. A.; Bergstrom, D. E. *Anal. Biochem.* **1984,** *139,* 243–262.

RECEIVED May 14, 1993

Chapter 11

Characterization of Processed Gag Proteins from Highly Replicating HIV-1MN

Catherine Fenselau[1], Xiaolan Yu[1], Duncan Bryant[1], Michelle A. Bowers[2], Raymond C. Sowder II[2], and Louis E. Henderson[2]

[1]Department of Chemistry and Biochemistry, University of Maryland Baltimore County, Baltimore, MD 21228
[2]AIDS Vaccine Program, Incorporated DynCorp, National Cancer Institute–Frederick Cancer Research and Development Center, Frederick, MD 21702

Mass Spectrometry techniques for both molecular weight determination and sequencing are applied to studies of mutation of the gag gene in cultured HIV-MN and of processing and posttranslational modification of the Gag proprotein. The techniques and strategies illustrated here can be used to provide definitive information on primary structures of proteins from all kinds of wild and chimeric microorganisms.

WHY STUDY VIRAL PROTEINS?

The most highly definitive molecules in microorganisms are usually considered to be the nucleic acids that contain genetic information. However, genetic material in the lente virus class undergoes rapid mutation, and in any given population many DNA variants will be available for cloning or elucidation via the polymerase chain reaction. Some of these variants will belong to specimens capable of replication, others will represent genetic dead ends. Consequently, if the objective is to characterize or define a viral population that has been successfully infecting and replicating, isolated protein may provide the more definitive fingerprint.

The MN strain of HIV-1 has been shown to be a prevalent prototype in the U.S.population, and standardized stocks of this strain have been prepared and characterized for use as challenge reagents in the development of antiviral strategies (1). The HIV genome contains three large open reading frames, called gag, pol and env, which direct synthesis of three large proproteins (2). These precursors are processed into smaller proteins that are incorporated into the maturing virus. Current understanding of the localizations of these products is schematically

0097–6156/94/0541–0159$06.00/0

illustrated in Figure 1. As indicated in that figure, one Gag-derived protein is a major structural component of the viral core, and the family carries out critical tasks during the assembly, budding, maturation and infection states of viral replication. The immature virion becomes infectious only after proteolytic cleavage of the Gag precursor and structural association of the products (3).Point mutations or small deletions in the gag gene can repress viral budding (4). In this context a study was undertaken to map and sequence the suite of proteins formed when the Gag proprotein is processed in maturing human immunodeficiency virus. The six products are the most abundant proteins readily found in the mature virus.

STRATEGY

The MN strain of human immunodeficiency virus type 1 was grown in cultured H9 cells. Cells were concentrated by centrifugation, disrupted and solubilized (1). Molecular components were fractionated by preparative reverse phase chromatography, using a very slow gradient. A portion of this chromatogram is shown in Figure 2, with peaks labelled that were found to contain the processed Gag proteins studied.

The strategy for characterizing these proteins and peptides was as follows. Molecular weights were determined for all peptides and proteins isolated. The large intact proteins (p17, p24 and p7) were weighed by electrospray mass spectrometry (5), while the smaller products of processing (p6, p2, and p1) were weighed by fast atom bombardment mass spectrometry (6). The molecular weights of the entire set might have been determined by laser desorption (7) or electrospray, whose mass ranges exceed the 30,000 daltons required for analysis of this set of proteins. However, fast atom bombardment using cesium ions as the bombarding particles easily provided accurate molecular ion determinations on p1 and p2, required for subsequent collisional activation to obtain sequence information (8,9).

The larger proteins were then digested with various endoproteinases to produce an array of peptide products, which were separated and isolated by reverse phase HPLC. These fractions were analysed or mapped by fast atom bombardment, which provided the molecular weights of the peptides. All protease products were known to be accounted for when the sum of their molecular weights equalled the molecular weight of the uncut protein (correcting for the elements of water added across each cleavage site).

The known proviral sequence of the gag gene (2) was used as the basis for computer simulated proteolytic digests (1) and the masses of the predicted peptides were compared with those obtained by fast atom bombardment. Conclusions could then be drawn about the extent of mutation and processing.

In addition to molecular weights, the masses of ionic fragments can be determined by mass spectrometry. Fragmentation of molecular ions is enhanced in a reproduceable manner in tandem mass spectrometry experiments in which protonated peptides undergo high energy gas phase collisions with helium atoms (8,9). Excellent mass accuracy is provided, both for selection of precursor ions, and for detection of fragment ions, in the four sector tandem instrument used in the present work. In a rare exception to Murphy's law, the bonds that break most readily in collisional dissociation are those that link togeher the amino acids to

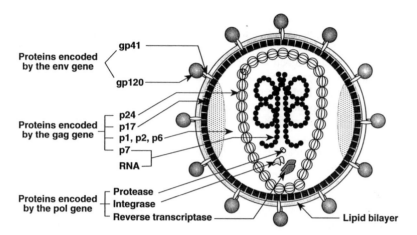

Figure 1. Schematic presentation of the structural organization of the HIV-1 virus and the locations of processed proteins from the three large proproteins coded by the genome. See text for more discussion of the processed Gag proprotein products.

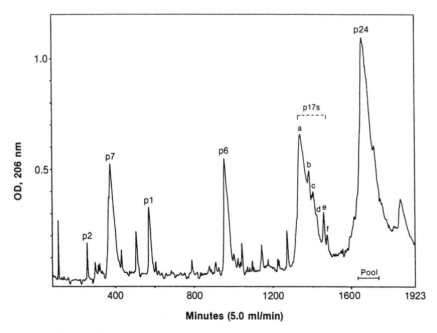

Figure 2. Partial reverse phase high pressure liquid chromatogram of disrupted and solubilized cultured viruses. This Figure is reproduced with permission from (1).

form the polymer. These bond cleavages are shown in Figure 3, along with the designations used in predicting and intepreting spectra. Because the twenty common amino acids have eighteen different masses, the presence and position of a residue may be inferred by the difference in the masses, for example of the a_2 and a_3 ions. The two pairs of amino acids with redundant masses, lysine and glutamine, and leucine and isoleucine, can be distinguished by additional fragmentation produced in the side chains in high energy collisional activation (8,9).

Whenever possible the amino acid sequences of the HPLC purified proteolytic peptides were obtained by both high energy collisional activation and automated Edman degradation. Based on the results of these complementary experiments, residue mutations and post-translational modifications could be located and characterized.

MOLECULAR WEIGHTS

Table I summarizes the molecular weights determined for peptides and proteins separated in the chromatogram shown in Figure 2. The dimer was found when p1 was rechromatographed and is probably artifactual. The first four values in the Table were determined by fast atom

Table I
Molecular Weights Determined for Processed Gag Proteins

Protein	Found Molecular Weight (daltons)	Calculated Molecular Weight
p1	1807.2 ±0.5	1806.9
p1 dimer	3610.8 ±1.0	3611.8
p6	6399.6 ±1.0	6400.6
p7	6450.8 ±1.0	6451.6
p17a	15269.3 ±3.5	15156.3
p24	25752 ±50.	25551.

bombardment mass spectrometry on a high resolution double focussing instrument. A partial FAB spectrum of p6 is shown in Figure 4. The last three molecular weights were determined by electrospray ionization on a quadrupole mass spectrometer. The first two masses in Table I are measured and calculated as monoisotopic masses (10), and the rest of the entries are average masses. Calculated values are based on the amino acid sequences determined in this study.

As has been reported (1), both p17 and p24 proteins were found to occur as variants. Multiple peaks in Figure 2 are marked as members of the p17 family. Five of these peaks were collected and at least partially characterized by protease mapping and sequencing, using mass spectrometry, HPLC and Edman degradation (1). Multiple variants were detected in each peak. The found and calculated molecular weights entered for p17a in Table I are for the major variant which consititutes about 50 % of all the p17 variants present in this infectious viral

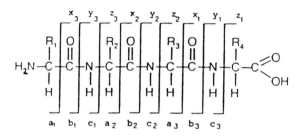

Figure 3. Schematic presentation of the six sequence ion series that can be formed by bond cleavages in the amide backbone of peptides. Notation follows that of Biemann (9).

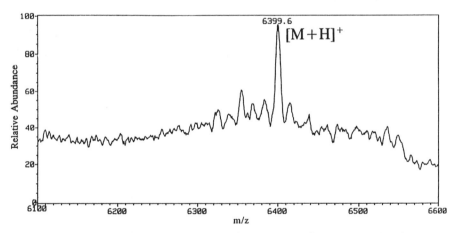

Figure 4. Molecular ions of protonated p6 analysed by fast atom bombardment on a double focussing mass spectrometer.

population. The difference between the found and calculated values reflects post-translational modification of the amino terminus by myristic acid, which interfered with sequencing by Edman degradation.

The upper mass range of the electrospray spectrum of p17a is shown in Figure 5. This technique generates suites of molecular ions carrying multiple charges (5), identified as +11 through +16 in Figure 5. The charge states (n) of protonated ions represented by adjacent peaks can be deduced from the m/z values (m), according to the formula $n_2 = (m_1 - 1)/(m_1 - m_2)$, where $n_2 = n_1 + 1$ (11). The molecular weight (M) can be calculated according to the formula $m_2 = (M + n_2)/n_2$ (11). This kind of transformation provided a molecular weight for p17a of 15269.3 $\pm$ 3.5 daltons.

The calculated value listed for p24 in Table I is for the sequence variant with alanine at position 86. The mass of a second variant found, with valine at position 86, is calculated to be 25579 daltons. Figure 6 presents the electrospray spectrum of this sample. Examination of the peaks in each charge state strongly suggested that this protein is heterogeneous. The average mass, calculated from the most intense peak in each charge state, is about 200 daltons heavier than either theoretical mass, reflecting post-translational modification thought to be a mixture of mono-, di- and tri-phosphorylation.

The found and calculated molecular weights listed in Table I for p7 are for denatured sample, and comfirm the absence of covalent post-translational modifications in this small protein. This protein has been found to form noncovalent complexes with divalent zinc and cadmium ions (12), which have been proposed to stabilize the conformation required for binding DNA. Such noncovalent complexes may also be characterized by mass spectrometry when care is taken in sample handling (13). Electrospray mass spectra are shown in Figure 7 of both (a) the denatured protein, and (b) a complex with two zinc cations. Deconvolution of the multiply charged ions in the spectrum in Figure 7b provided a molecular weight of 6582.0, divergent from the molecular weight deduced for the denatured protein by a mass difference that corresponds to two zinc ions.

PROTEIN MAPS AND SEQUENCES

Based on combinations of Edman degradation and mass spectrometry, the amino acid sequences of p2, p6 and p7 were found to be identical to those predicted by the sequence of proviral cDNA. Five discrepancies were found in p17a variants, five in p24 and two in p1 (1). Sequence heterogeneity was found mainly in the p17 protein, and one heterogeneous site was characterized in p24. The complex situation for p17 is summarized in Figure 8. At least eight different sequence variants have been recognised from the four HPLC peaks containing p17a,b,c and d proteins.

The strategy by which this sequence information was obtained can be illustrated with the p17 variants. After preparative scale reverse phase HPLC, the individual variants were digested with S.aureus V8 and chymotrypsin. The peptide products of each digest were fractionated by reverse phase HPLC. An example is shown in Figure 9, in which products of digestion with S. aureus V8 are identified (1). The HPLC peaks were collected as fractions, lyophilized, and the peptides

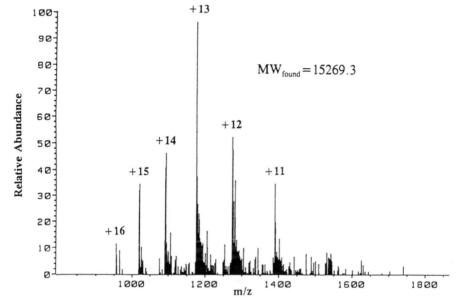

Figure 5. Molecular ions formed from p17a by electrospray ionization on a quadrupole mass spectrometer. Deconvolution of the suite of multiply charged molecular ions to obtain the molecular weight is discussed in the text. This Figure is reproduced with permission from (1).

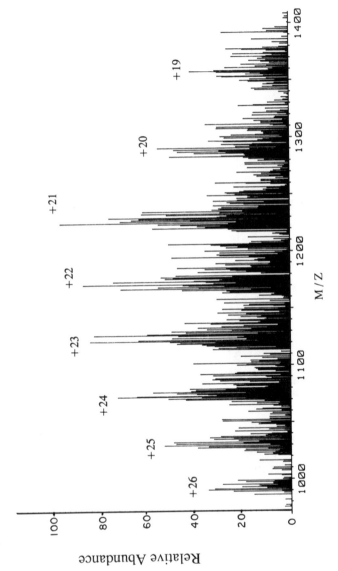

Figure 6. Molecular ions formed from p24 by electrospray ionization on a quadrupole mass spectrometer. Although the charge state could be assigned, calculation of the moleculer weight was less precise, because the sample is heterogeneously phosphorylated.

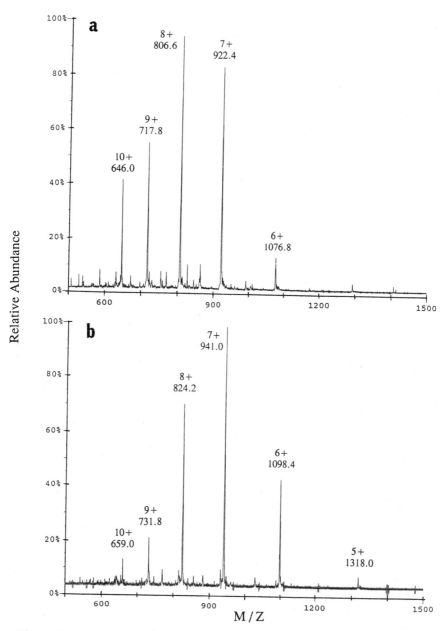

Figure 7. Molecular ions formed from p7 by electrospray ionization. a) protein denatured; b) protein complexed with two divalent zinc cations.

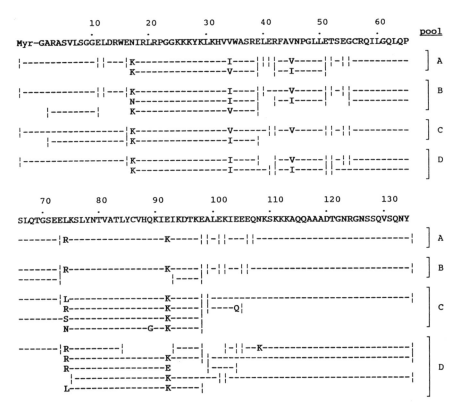

Figure 8. Amino acid sequences determined for variants of p17. On the right side A,B,C and D refer to protein pools separated as indicated in Figure 2. The amino acid sequence predicted by the proviral DNA sequence is shown in heavy type on the top line. Dashes indicate regions or positions in which the predicted and determined sequences agree. This Figure is reproduced with permission from (1).

contained in the fractions were sequenced (where possible) by Edman degradataion and tandem mass spectrometry using collisional activation. For success in the latter case, the peptide had to be efficiently desorbed by fast atom bombardment and short enough (<2500 Da) to produce structurally informative fragment ions.

One advantage of tandem mass spectrometry is its ability to characterize peptides with derivatized amino termini. Figure 10 shows the collisional activation mass spectrum of the myristylated amino terminal peptide [1-11] collected as a late running (hydrophobic) peak in the chromatogram in Figure 9. Interpretation of the spectrum clearly localizes the modification on terminal glycine. Collisional activation spectra are shown in Figure 11 of the heterogeneous pair of peptides [42-51] recovered in two separate peaks in Figure 9. Both the molecular weights and the fragmentation patterns (illustrated in the Figure for 'a' series ions) define valine at position 45 in the faster eluting peptide and isoleucine as residue 45 in the peptide in the second HPLC peak.

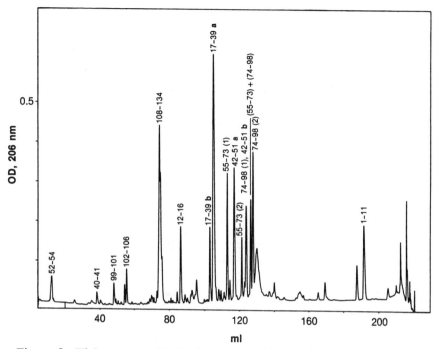

Figure 9. High pressure liquid chromatographic separation of the peptide mixture produced by incubation of p17a with S.aureus V8 protease. This Figure is reproduced with permission from (1).

CONCLUSIONS

The structural heterogeneity defined by direct sequencing of proteins isolated from a single production lot of HIV-1MN probably results from the expression of several different proviruses. The most replicative (infectious) will be expanded most successfully during cell culture. Thus the proteins sequenced are characteristic of the more replicative viruses and show the sequences of Gag precursors that function efficiently in the assembly and budding processes.

Viral proteins have also been sequenced by several other mass spectrometry laboratories (14-16). The relevance of the characterization of post-translational modifications in viral proteins to the development of alternate therapeutic agents, diagnostic reagents and vaccines for viral infections has been emphasized recently by Gorman (14).

The methods illustrated here can be applied to characterize other viruses produced by cell culture, and potentially, to distinguish strains. Mass spectrometry can provide definitive structural information on proteins from other microorganism populations (wild and chimeric) to complement studies of biological function (17-19).

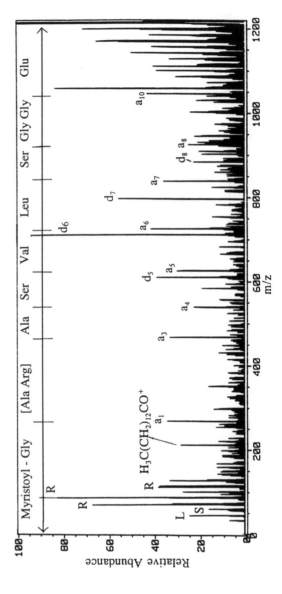

Figure 10. Mass spectrum obtained using high energy collisional activation on a four sector tandem mass spectrometer of the peptide eluting in the peak labelled 1-11 in Figure 8. The amino acids shown in sequence are deduced from the series of 'a' ions in the spectrum. This Figure is reproduced with permission from (8). Notation follows that of Biemann (9).

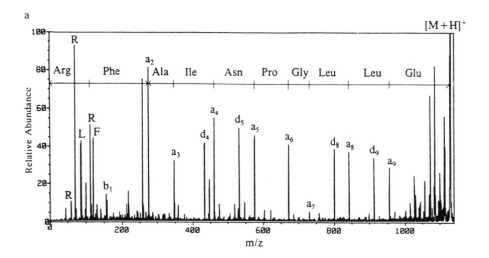

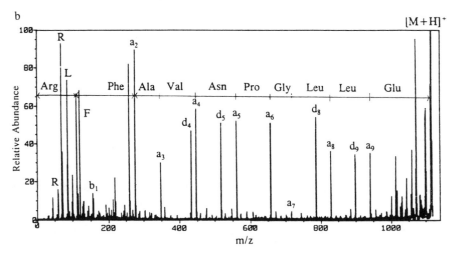

Figure 11. Mass spectra obtained using high energy collisional activation on a four sector tandem mass spectrometer of homologous peptides eluting in the two peaks labelled 42-51 in Figure 8. The amino acids shown in sequence are deduced from the series of 'a' ions in the spectrum. This Figure is reproduced with permission from (1).

ACKNOWLEDGMENTS

Research at UMBC was supported by grants from the National Science Foundation (DIR 14549) and the National Institutes of Health (GM 21248). Research at FCRDC was supported by the National Cancer Institute under contract NO1-CO-74102 with Program Resources, Incorporated/DynCorp.

LITERATURE CITED

1. Henderson, L.E.; Bowers, M.A.; Sowder, R.C.; Serabyn, S.A.; Johnson, D.G.; Bess, J.W.; Arthur, L.O., Bryant, D.K.; Fenselau, C. J.Virol. 1992, 66, 1856.
2. Myers, G.; Josephs, S.F.; Rabson, A.B.; Smith, T.F.; Wong-Staal, F.,Eds.; Human Retroviruses and AIDS 1988, a Compilation and Analysis of Nucleic Acid and Amino Acid Sequences; Los Alamos National Laboratory, Los Alamos, NM, 1988.
3. Kohl, N.E.; Emini, E.A.; Schleif, W.E.; Davis, L.J., Heimbach, J.C.; Dixon, R.A.; Scolnick, E.M.; Sigal, I.S. Proc. Nat. Acad. Sci. USA 1988, 85, 4686.
4. Trono, D.; Feinberg, M.B.; Baltimore, D. Cell 1989, 59, 113.
5. Smith, R.D.; Loo, J.A.; Loo, R.R.; Busman, M.; Udseth, H.R. Mass Spectrom. Reviews 1991, 10, 359.
6. Fenselau, C.; Cotter, R.J. Chem. Rev. 1987, 57, 501.
7. Hillenkamp, F.; Karas, M.; Beavis, R.C.; Chait, B.T. Anal. Chem. 1991, 63, 1193a.
8. Fenselau, C. Annu. Rev. Biophys. Biophys. Chem. 1991, 20, 205.
9. Biemann, K. Annu. Rev. Biochem. 1992, 61, 977.
10. Yergey, J.; Heller, D.; Hansen, G.; Cotter, R.J.; Fenselau, C. Anal.Chem. 1983, 55, 353.
11. Mann, M.; Meng, C. K.; Fenn, J. B. Anal.Chem. 1989, 61, 1702.
12. Summers,M.F.; Henderson, L.E.; Chance, M.R.; Bess, J.W.; South, T.L.; Blake, P.R.; Sagi, I.; Perez-Alvarado, G.; Sowder, R.C.; Hare, D.R.; Arthur, L.O. Protein Sci. 1992, 1, 563.
13. Yu, X.; Fenselau, C. Anal.Chem. 1993, 65, 1355.
14. Gorman, J.J. Trends in Analytical Chemistry 1992, 11, 96.
15. Gorman, J.J.; Corino, G.L.; Shiell, B.J. Biomed. Environ. Mass Spectrom. 1990, 19, 646.
16. Chait, B.T.; Kent, S.B. Science 1992, 257, 1885.
17. Carr, S.A.; Hemling, M.E.; Bean, M.F.; Roberts, G.D. Anal.Chem. 1991, 63, 2802.
18. Bockerhoff, S.E.; Edmonds, C.G.; Davis, T.N. Protein Science 1992, 1, 504
19. Vestling, M.; Murphy, C.; Fenselau, C.; Chen, T.T. Molec. Marine Biol. Biotech. 1991, 1, 73.

RECEIVED July 8, 1993

Chapter 12

Mass Spectral Analysis of Lipid A of Gram-Negative Bacteria

K. Takayama[1,2], N. Qureshi[1,2], and R. J. Cotter[3]

[1]Mycobacteriology Research Laboratory, William S. Middleton Memorial Veterans Hospital, Madison, WI 53705
[2]Department of Bacteriology, College of Agricultural and Life Sciences, University of Wisconsin, Madison, WI 53706
[3]Department of Pharmacology, Johns Hopkins University School of Medicine, Baltimore, MD 21205

Positive ion FAB-MS, LD-MS, and PD-MS were utilized to obtain important structural information on the 4'-monophosphoryl lipid A (MPLA) derived from the lipopolysaccharides of *Salmonella typhimurium* and other Gram-negative bacterial strains which led to the elucidation of its chemical structure. The spectra of MPLA generated by these methods showed peaks corresponding to cationized molecular ions, losses of fatty acyl groups, and cleavage of the disaccharide yielding oxonium ions (FAB-MS and PD-MS), and several two-bond ring cleavages (LD-MS and PD-MS). The distribution of the fatty acyl groups on the MPLA molecule was then deduced from its size and fragmentation pattern. We suggest that one could now identify and differentiate certain Gram-negative bacterial strains by performing mass spectral analysis of the isolated MPLA.

Gram-negative bacteria contain characteristic amphipathic glycolipids on their cell surface called lipopolysaccharides (LPS) (*1*). LPS isolated from the smooth strains of *Enterobacteriaceae* are composed of three structural regions: the hydrophilic 0-specific polysaccharide, the common core oligosaccharide, and lipid A which anchors the LPS to the outer membrane by hydrophobic effect (*2*). The lipid A derived from the LPS of *Escherichia coli* and *Salmonella* strains is a 1,4'-diphosphorylated glucosamine disaccharide containing fatty acyl groups which are designated as R_1 to R_4 as shown in Figure 1. In the hexaacyl lipid A, R_1 is myristoxymyristate, R_2 is lauroxymyristate, and R_3 and R_4 are hydroxymyristates.

The lipid A moiety of LPS is biologically active and is responsible for many of the pathophysiological effects associated with Gram-negative bacterial infections. These effects include shock, pyrogenicity, hemodynamic changes, and disseminating intravascular coagulation (*3*). The structure of this highly conserved molecule was elucidated in 1983, and one of the decisive instrumental techniques responsible for this was fast atom bombardment mass spectrometry (FAB-MS) (*4–6*). Further

0097–6156/94/0541–0173$06.00/0
© 1994 American Chemical Society

advances in mass spectral analysis of lipid A were made by use of infrared laser desorption mass spectrometry (LD-MS) and ^{252}Cf plasma desorption mass spectrometry (PD-MS) (7–10). By using PD-MS, it was possible to examine the structures of intact LPS from the deep rough mutant of *E. coli* (ReLPS) (9).

We shall discuss how the three types of "desorption" methods have been used to examine the structures of LPS and lipid A and how such methods might be applied in identifying Gram-negative bacteria.

FAB-MS Analysis of Lipid A

Positive ion FAB-MS was used for the first time on 4′-monophosphoryl lipid A (MPLA) derived from the LPS of *Salmonella typhimurium* by Qureshi et al. in 1982 (11). In these early studies, only the protonated molecular ions (MH$^+$) at m/z 1716, 1507, and 1281 representing hexaacyl-, pentaacyl-, and tetraacyl-MPLA, respectively, could be discerned for the silica gel thin layer chromatography–purified structural series. These results allowed one to determine the molecular formula and confirm the probable chemical composition of the lipid A. Improvements in the analysis of lipid A (about 20 μg) by FAB-MS was reported by Qureshi et al. in 1983 (4). In this study, a key fragmentation of lipid A was identified as being the formation of an oxonium ion at m/z 1087 and 876. It involved the cleavage of the glycosidic bond and the formation of a positively charged distal subunit as shown in Figure 2. This allowed for the first time the determination of the distribution of the fatty acyl groups in the glucosamine disaccharide backbone. This was one of the key instrumental techniques used that led to the final elucidation of the structure of the lipid A from the LPS of *S. typhimurium* in 1983 by Takayama et al. (5) (see Table I).

Lipid A was derivatized by methylation (at the phosphate with diazomethane), fractionated by reverse-phase high performance liquid chromatography (HPLC) and analyzed by positive ion FAB-MS. These results showed not only the presence of molecular ions (MH$^+$, MNH$_4$$^+$, and MNa$^+$) and an oxonium ion but also molecular ions minus ester-linked fatty acyl groups (6). The cleavage of the ester-linked fatty acyl groups occurred on either side of the connecting oxygen atom. When the cleavage was between the oxygen and the sugar, an additional hydrogen was lost to produce an even electron ion, which introduced a double bond in the sugar ring. When the cleavage occurred between the oxygen and carbonyl function of the fatty acyl group, a hydrogen was transferred to the sugar to produce an even electron ion with a hydroxyl group at that position. Thus two pathways resulted in fragment ions that were 18 amu apart. FAB-MS is now adequately developed to determine the structure of lipid A.

LD-MS of Lipid A

Seydel et al. (13) were the first to utilized LD-MS (positive ion mode) for the structural analysis of lipid A. Since they encountered difficulties in obtaining interpretable spectra and molecular ion peaks of lipid A, possibly due to high polarity, they had to dephosphorylate their samples (14). Thus, they examined the

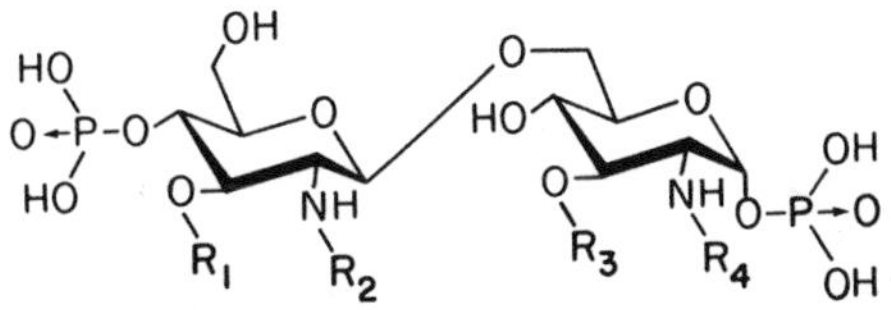

Figure 1. General structure of DPLA derived from Gram-negative bacteria. In this structure, the sugar is glucosamine. R_1 to R_4 represent the fatty acyl groups. DPLA is obtained by hydrolyzing LPS in 0.02 M sodium acetate buffer, pH 4.5, at 100°C for 60 min, whereas MPLA is obtained by hydrolysis in 0.1 M HCl, 100°C for 15-30 min (12).

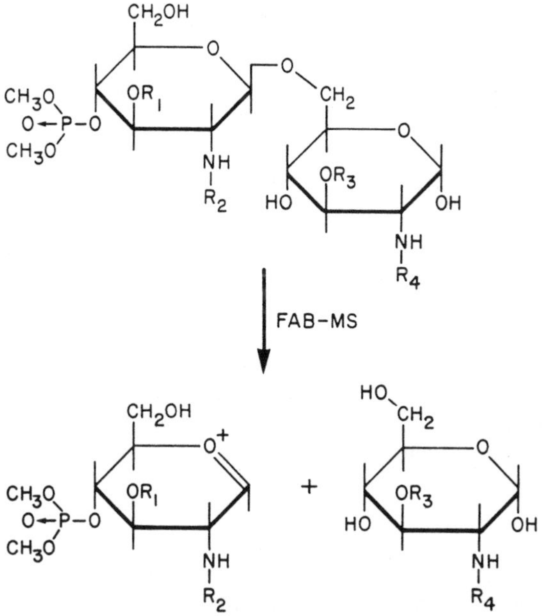

Figure 2. Formation of oxonium ion by the cleavage of the glycosidic bond of dimethyl MPLA by positive ion FAB-MS and PD-MS. R_1 to R_4 represent the fatty acyl groups.

Table I. Structure of Lipid A Derived from Select Gram-negative Bacterial Sources[a]

Lipid A Source[c]	No. of Fatty Acids	Fatty Acyl Group[b]			
		R_1	R_2	R_3	R_4
Escherichia coli[d]	6	$C_{14}OC_{14}$	$C_{12}OC_{14}$	OHC_{14}	OHC_{14}
Salmonella typhimurium[d]	6	$C_{14}OC_{14}$	$C_{12}OC_{14}$	OHC_{14}	OHC_{14}
Salmonella minnesota[d]	7	$C_{14}OC_{14}$	$C_{12}OC_{14}$	OHC_{14}	$C_{16}OC_{14}$
	6	$C_{14}OC_{14}$	$C_{12}OC_{14}$	OHC_{14}	OHC_{14}
Haemophilus influenzae[e]	6	$C_{14}OC_{14}$	$C_{14}OC_{14}$	OHC_{14}	OHC_{14}
Proteus mirabilis[d]	7	$C_{14}OC_{14}$	$C_{14}OC_{14}$	OHC_{14}	$C_{16}OC_{14}$
	6	$C_{14}OC_{14}$	$C_{14}OC_{14}$	OHC_{14}	OHC_{14}
Shigella sonnei[d]	6	OHC_{14}	$C_{14}OC_{14}$	OHC_{14}	$C_{12}OC_{14}$
Neisseria gonorrhoeae[f]	6	OHC_{12}	$C_{12}OC_{14}$	OHC_{12}	$C_{12}OC_{14}$
	5	OHC_{12}	$C_{12}OC_{14}$	OHC_{12}	OHC_{14}
Pseudomonas aeruginosa[g]	6	OHC_{10}	$C_{12}OC_{12}$	OHC_{10}	$C_{12}OC_{12}$

[a]Taken from ref. 12.

[b]Refer to Figure 1 for the structure of DPLA and location of the fatty acyl groups.

[c]Bacterial classification according to Bergy's Manual of Systemic Bacteriology, 1984: Facultatively Anaerobic Gram-Negative Rods, [d]Family I. *Enterobacteriaceae* and [e]Family III. *Pasteurellaceae*; Gram-Negative Aerobic Rods and Cocci, [f]Family VIII. *Neisseriaceae* and [g]Family I. *Pseudomonadaceae*.

structure of dephosphorylated lipid A derived from the LPS of *Salmonella minnesota, E. coli,* and *Proteus mirabilis* by LD-MS (*13*). The spectra contained molecular ion peaks MK^+ and several fragment ion peaks including those for MK^+ minus ester-linked fatty acyl groups as well as the cation of the monosaccharide subunits similar to the oxonium ion. Cleavage of the glycosidic bond appeared to occur on both sides of the connecting oxygen. Since the two monosaccharide subunits were structurally equivalent in the dephosphorylated samples, it was not possible to directly identify the fragments. Tabet and Cotter (*15*) performed the LD-MS analysis of MPLA derived from the LPS of *S. typhimurium* and showed the presence of a protonated molecular ion. This suggested that it is possible to analyze lipid A by LD-MS directly without dephosphorylation.

LD-MS was used by Takayama et al. (7) to characterize the structural series of MPLA derived from the LPS of *Neisseria gonorrhoeae*. Fractionation of MPLA by HPLC required methylation of the 4'-phosphate group, which reduced the polarity and made it compatible with the use of laser desorption. Two pentaacyl (M_r 1463 and 1479) and two hexaacyl (M_r 1661 and 1677) MPLA were identified by LD-MS and confirmed by FAB-MS. The major hexaacyl MPLA component had R_1 and R_3 = hydroxylaurate and R_2 and R_4 = lauroxymyristate (M_r 1661), whereas the major pentaacyl MPLA component had R_1 and R_3 = hydroxylaurate, R_2 = lauroxymyristate, and R_4 = hydroxymyristate (M_r 1479) (refer to Figure 1 and Table I). The spectra were characterized by the presence of molecular ion peaks MK^+ and fragment ion peaks minus either hydroxylauroyl or lauroyl group.

Significantly, two-bond ring cleavages of the reducing-end sugar were observed in LD-MS (Figure 3). These peaks were at m/z 974, 1034, and 1233, which represented three major fragmentations. Cleavages of the C_1-O and C_2-C_3 bonds of the reducing-end sugar (cleavage A) resulted in the loss of a portion of the molecule containing the amide-linked fatty acid. If the amide-linked fatty acid (R_4) was lauroxymyristate, then the loss of 467 amu produced a peak at m/z 1233. A second two-bond ring cleavage (cleavage B), which is similar to the cleavage A, included a loss of both amide-linked (R_4) and ester-linked (R_3) fatty acids. This was used to distinguish lauric or myristic acid from the hydroxylauric acid at the 3 position if R_4 was known. A loss of 665 amu produced a peak at m/z 1035, which established R_3 as hydroxylaurate. The third fragmentation involved the cleavage of the C_1-O and C_4-C_5 bonds (cleavage C). When R_1 and R_2 were hydroxylaurate and lauroxymyristate, respectively, an ion at m/z 974 would result. Spengler et al. (16) proposed a mechanism for cleavages A and C. They suggested that the hemiacetal ring of the reducing-end sugar first opens due to rapid heating by infrared laser irradiation followed by several retro-aldol-type reactions. They also suggested that there is a final ester pyrolysis reaction which eliminates R_3 and yields a fragment that is equivalent to cleavage B as shown in Figure 3.

The tetramethyl diphosphoryl lipid A (DPLA) derived from the LPS of *E. coli* contains a dimethyl phosphate group at the anomeric carbon and at the C_4 of the distal sugar. The reducing-end phosphate was easily cleaved from MK^+ (1893 amu) to yield three types of fragments. Cleavage occurred on either side of the oxygen that connects the phosphate group to the sugar ring, and it may have involved either proton or methyl transfer. The loss of 94, 108, and 126 amu would correspond to the loss of PO_3CH_3, $PO_3(CH_3)(CH_2)$, and $HPO_4(CH_3)_2$, respectively. Spengler et al. (16) have suggested that the loss of the reducing-end phosphate is accompanied by the opening of the hemiacetal ring. LD-MS was thus developed into an effective method to determine the structure of lipid A as the methylated derivative. It was used successfully to complete the structural determination of the HPLC-purified dimethyl MPLA derived from the nontoxic LPS of *Rhodobacter sphaeroides* (17). This and further studies by Qureshi et al. (18) showed that fine-structural differences exist between a toxic and a nontoxic DPLA, the former being derived from the LPS of *E. coli*. The lipid A from *R. sphaeroides* contained: (a) only five fatty acids as compared to six for the toxic lipid A; (b) short chain hydroxy fatty acids

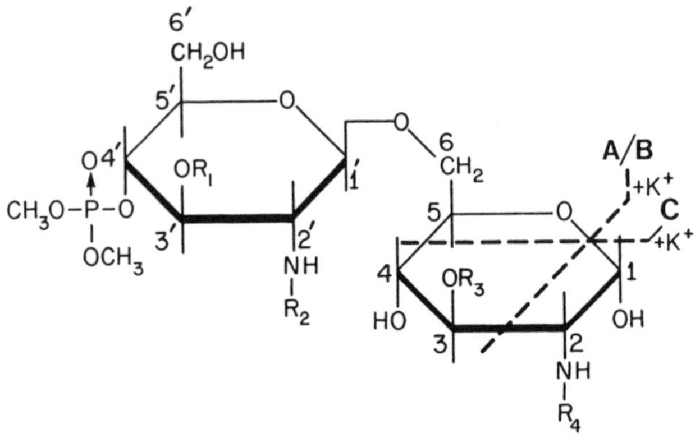

Figure 3. Two-bond ring cleavages by the fragmentation of the reducing-end subunit of dimethyl MPLA by positive ion FAB-MS and PD-MS. These cleavages are indicated by dotted lines and are labeled A, B, and C. Cleavage B also involves the loss of the OR_3 group. Only cleavage C is seen in PD-MS. R_1 to R_4 represent the fatty acyl groups.

(hydroxycapric vs. hydroxymyristic acid in the toxic lipid A); and (c) greater variation in the chain-length of the fatty acids.

PD-MS Analysis of Lipid A

PD-MS was used for the first time for the structural analysis of the ReLPS from E. coli (9) and later for analysis of the lipid A derived from the LPS of E. coli and R. sphaeroides (10). Several forms of lipid A were examined, including MPLA and DPLA and their methyl derivatives.

Lipid A Derived from LPS of E. coli. The free acid and the methyl ester forms of this lipid A were examined by PD-MS to determine their ability to form molecular ions under heavy bombardment and to determine the kinds of fragment ions produced to yield structural information (10).

Molecular Ion. DPLA would be expected to be more labile than the MPLA because of the presence of the labile glycosidic phosphate. However, by PD-MS, both forms of E. coli lipid As were adequately stable to yield molecular ions (data not presented). The molecular ion regions of the spectra of MPLA are shown in Figure 4 (10). In the positive ion mode, both free acid and the methyl derivatives gave intense MNa^+ at m/z 1741 and 1769, respectively. The free acid also formed MH^+ ion at m/z 1719 and a species containing two sodium ions $(M - H + 2Na)^+$ at m/z 1763. The methyl derivative formed $(M - CH_2 + Na)^+$ ion at m/z 1755 that most likely reflects incomplete methylation. In the negative ion mode, the free acid form gave $(M - H)^-$ ion at m/z 1717 and $(M - CH_3)^-$ ion at m/z 1703, whereas the methyl derivative gave MCl^- ion at m/z 1781 and $(M - CH_3)^-$ ion at m/z 1731. The signal to noise ratio of the positively charged molecular ion peaks were better than that of the negatively charged molecular ions. The free acid form gave a better molecular ion signal in the negative ion mode than the methyl derivative.

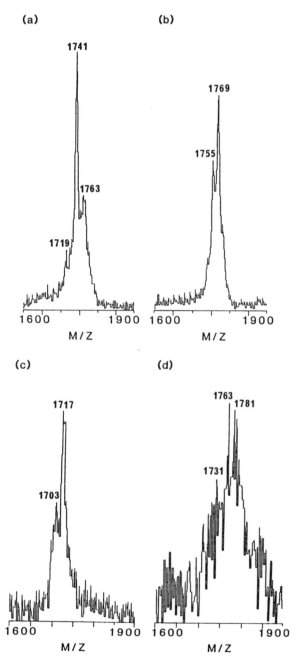

Figure 4.   Molecular ion regions of PD-mass spectra of *E. coli* MPLA: (a) Positive ion mass spectrum of free acid, (b) positive ion mass spectrum of the methyl derivative, (c) negative ion mass spectrum of free acid, and (d) negative ion mass spectrum of the methyl derivative. (Reproduced with permission from ref. 10. Copyright 1992 Elsevier Science Publishers B.V.)

High Mass Fragments. The free acid form of MPLA was examined in both positive and negative ion modes of PD-MS (Figure 5) (*10*). The predominant feature of this region was the loss of the ester-linked fatty acyl groups (myristoxymyristoyl, hydroxymyristoyl, and lauroyl groups). In Figure 5a, the loss of 243 and 227 amu from MNa$^+$ (m/z 1741) to yield ions at m/z 1498 and 1514, respectively, are due to loss of hydroxymyristate by cleavage at either side of the oxygen of the ester. We have found that the cleavage of the fatty acyl side of the ester oxygen is prevalent in PD-MS. A loss of 437 amu yielded an ion at m/z 1304 which was a loss of the myristoxymyristoyl group. In the negative ion mode (Figure 5b), the peak at m/z 1717 corresponds to the loss of hydrogen to form a (M - H)$^-$ ion. Peaks at m/z 1490, 1280, and 1051 correspond to the loss of hydroxymyristoyl, myristoxymyristol, and both hydroxymyristoyl and myristoxymyristoyl moieties, respectively. Negative ion PD-MS was shown to selectively cleave the fatty acyl esters linked directly to the sugar, whereas positive ion PD-MS cleaved both sides of the oxygen linkage. In the positive ion PD-MS, an oxonium ion (at m/z 1088) was formed by the cleavage of the glycosidic bond between the distal and reducing-end sugars.

Low Mass Fragments. The spectra of negative ion PD-MS showed the presence of fatty acids and phosphate anions (*10*). Ester bonds were cleaved to yield the carboxylate anions for laurate (m/z 199), myristate (m/z 227), and hydroxymyristate (m/z 243). Fragment ions at m/z 79 and 97 were the PO$_3^-$ and H$_2$PO$_4^-$ anions, respectively. In dimethyl MPLA and tetramethyl DPLA, the presence of the methylated phosphate group was determined by the presence of (CH$_3$)HPO$_4^-$ and (CH$_3$)$_2$HPO$_4^-$ anions at m/z 111 and 125, respectively. Negative ion PD-MS was found to be superior to positive ion PD-MS in obtaining structural information in the low mass range.

Effect of Methylation on Fragmentation of Lipid A. The intensity of the fragment ion peaks was slightly lower for the methylated MPLA than the free acid (*10*). This might be due to changes in conformation and electron distribution caused by the methyl groups on the phosphate. In the negative ion mode, the same fragmentation patterns were observed with free acids and the methyl derivative of MPLA. However, the spectrum of the methylated derivative was more difficult to interpret.

Formation of Both Oxonium and Two-bond Ring Cleavage Ions. PD-MS of both dimethyl MPLA and tetramethyl DPLA derived from the LPS of *R. sphaeroides* revealed the presence of an oxonium ion peak at m/z 875 and a two-bond ring cleavage ion peak at m/z 957 (cleavage C) (*10*). Thus, the two types of fragmentation of the disaccharide previously shown in FAB-MS (*4*) and LD-MS (*7*) are found in PD-MS.

Effect of 2-Keto-3-Deoxyoctonate (Kdo) on Fragmentation of Lipid A. The hexamethyl ReLPS from *E. coli* D31m4 which contains two Kdo units was analyzed by PD-MS (9). Several molecular ions were observed suggesting a variable degree of methylation. MNa$^+$ ions at m/z 2360, 2373, and 2387 containing additional

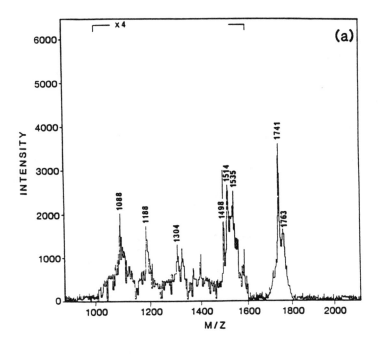

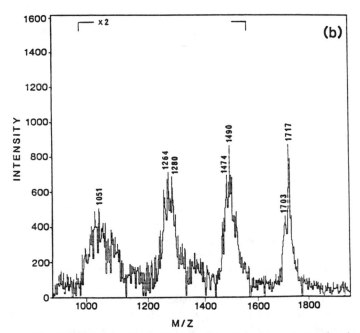

Figure 5.   The high mass regions of the positive ion (a) and negative ion (b) PD-mass spectra of the free acid form of *E. coli* MPLA.   (Reproduced with permission from ref. 10.   Copyright 1992 Elsevier Science Publishers B.V.)

CH_2, $2CH_2$, and $3CH_2$ groups, respectively, were observed.   The major fragmentation ion was the MNa^+ minus dimethyl phosphate from the reducing end. The loss of the hydroxymyristoyl group from the 3-position of the lipid A (R_3) was also observed.   The presence of the two Kdo units did not influence these fragmentations, except for the apparent inhibition of cleavage of the ester-linked myristoxymyristoyl group at the 3'-position (R_1).

Identification of Gram-negative Bacteria by Mass Spectrometry

Identification Based on Structure of Lipid A. The structure of the lipid A moiety of LPS from Gram-negative bacteria is highly conserved.   The disaccharide backbone is composed of either glucosamine (which is common), 2,3-diamino-2,3-dideoxy-glucose (DAG), or a mixture of the two components (12). The disaccharide is usually phosphorylated at the 1- and 4'-positions, and hydroxy fatty acids are directly linked to the sugar in either amide or ester linkage at the 2-, 3-, 2'-, and 3'-positions (Figure 1).   These hydroxy fatty acids may also carry straight-chain, nonhydroxy fatty acids in ester linkage (acyloxyacyl group).   There is considerable variation in the composition and distribution of the fatty acyl groups in lipid A, depending on the bacterial source of the LPS (12). This might serve as the basis for identifying Gram-negative bacteria.

In such an analysis, the LPS would first have to be extracted from the Gram-negative bacteria by the hot phenol–water method (19) for the smooth LPS or by the chloroform–phenol–petroleum ether–water method (20) with modifications (4) for the rough LPS.   The lipid A would then be prepared by mild acid hydrolysis of the LPS to yield either the MPLA or DPLA (3).   The free acid or methylated form of this lipid A can be analyzed by soft ion mass spectrometry.

Based on structure, the major lipid A components derived from the LPS of E. coli and the Salmonella strains (from the same family) are essentially identical (Table I). Because of the unequal distribution of the fatty acids within the structure, they are considered to be the asymmetrical hexaacyl lipid A (12). Differentiation between these two strains can be made by showing that the polar substituents of 4-aminoarabinose and phosphorylethanolamine attached to the two phosphates of the lipid A moiety (see Figure 1) are present in the LPS (12). The LPS from E. coli lacks these polar substituents on the lipid A.   Evidence for the presence of these polar substituents in the lipid A can be obtained by performing both negative and positive ion FAB-MS (21).   However, since the preparation of lipid A containing these labile groups from the LPS is difficult, chemical analysis of the LPS might be preferable.   The LPS of S. typhimurium and S. minnesota can be differentiated by examining the fatty acid composition and distribution.   The lipid A from the LPS of S. minnesota contains some heptaacyl lipid A in which the extra palmitic acid is ester-linked to the hydroxymyristic acid at the 2-position.   FAB-MS was used by Qureshi et al. to show this difference (22).

Although the LPS of Haemophilus influenzae, P. mirabilis, and Shigella sonnei contains only hydroxymyristic and myristic acids, they can be differentiated by their lipid A composition (Table I). It is to be noted that H. influenzae is from a different Family.   The six fatty acids in the lipid As of H. influenzae and P. mirabilis have asymmetrical distribution, whereas the fatty acids in the lipid A of *S. sonnei* have symmetrical distribution. This difference can be determined by the nature of the oxonium ion or two-bond ring cleavage ions in FAB-, LD-, and PD-MS. *H. influenzae* and *P. mirabilis* can be differentiated by the presence of heptaacyl lipid A in the latter case. This would be revealed by examining the molecular ions.

   Pseudomonas aeruginosa and *Neisseria gonorrhoeae* (from different Families) can be differentiated by analyzing their lipid A structures (Table I). The symmetrical hexaacyl lipid A of *P. aeruginosa* contains hydroxycapric, hydroxylauric, and lauric acids, whereas that of *N. gonorrhoeae* contains hydroxylauric, hydroxymyristic, and lauric acids. Thus, these two bacteria could be differentiated by determining the sizes (molecular ions) of their lipid As by mass spectrometry. The lipid A from the LPS of *N. gonorrhoeae* should be 112 amu larger than the corresponding lipid A from *P. aeruginosa*.

Lipid A Containing DAG Backbone. Takayama and Qureshi (*12*) listed 16 different Gram-negative bacteria whose lipid A moieties contain DAG instead of glucosamine. The complete structures of these lipid As have not been established. Qureshi et al. (manuscript in preparation) showed that the lipid A from the LPS of *Brucella abortus* has a DAG disaccharide backbone. The glycosidic bond of such a structure was found to be resistant to cleavage by FAB-MS. LD-MS of such a sample did not yield the two-bond ring cleavage. Such a structure was also characterized by the presence of stable amide-linked hydroxy fatty acyl groups. Only the acyloxyacyl fatty acyl groups were susceptible to cleavage by mass spectrometry. Thus FAB-, LD-, or PD-MS of these lipid As would yield only the molecular ions and the fragment ions MNa^+ minus nonhydroxy fatty acyl groups. These properties might be used to identify the DAG disaccharide-containing lipid As by mass spectrometry.

Conclusion

LD-MS appears to provide more structural information of lipid A than either FAB- or PD-MS. However, PD-MS can in addition be used for the structural analysis of ReLPS. Although soft ion mass spectrometry has been crucial in the determination of the structure of LPS and Lipid A, it is not possible to identify a Gram-negative bacteria based solely on this method. However, because of the observed variations even within strains, it is possible to differentiate certain bacteria based on the lipid A structure. This might be useful in cases where the identity is uncertain. This will become evident as many more of these structures from the LPS of Gram-negative bacteria are elucidated.

   The chemical structure of the lipid A moiety of LPS is important because almost all of the biological activities of LPS (toxicity and activation of the immune system in mammals) reside in this structure. Subtle differences in this structure can affect these activities. For example, a decrease in the number of fatty acyl groups from six to five or four, an increase in the number of fatty acyl groups from six to seven, a decrease in the size of the fatty acyl groups from C_{14} to C_{10}, and a decrease in the number of phosphate groups from two to one all lead to a significant decrease in LPS's biological activities (*3,12*). All of these structural differences can be easily differentiated by soft ion mass spectrometry.

Acknowledgments

This study was supported by the Research Service of the Department of Veterans Affairs and by the National Institutes of Health grant GM-36054.

Literature Cited

1. Galanos, C.; Lüderitz, O.; Rietschel, E. T.; Westphal O. *Int. Rev. Biochem.* **1977** *14*, 239–335.
2. Rietschel, E. T.; Galanos, C.; Lüderitz, O.; Westphal O. In *Immunopharmacology and Regulation of Leukocyte Function*; Webb, D. R., Ed.; Marcel Dekker: New York, NY, 1982; pp. 183–229.
3. Qureshi, N.; Takayama, K. In *The Bacteria. A Treatise on Structure and Function*; Iglewski, B. H.; Clark, V. L., Eds.; Academic Press, Inc.: San Diego, CA, 1990, Vol. XI; pp. 319–338.
4. Qureshi, N.; Takayama, K.; Heller, D.; Fenselau, C. *J. Biol. Chem.* **1983**, *258*, 12947–12951.
5. Takayama, K.; Qureshi, N.; Mascagni, P. *J. Biol. Chem.* **1983**, *258*, 12801–12803.
6. Qureshi, N.; Cotter, R. J.; Takayama, K. *J. Microbiol. Methods* **1986**, *5*, 65–77.
7. Takayama, K.; Qureshi, N.; Hyver, K.; Honovich, J.; Mascagni, P.; Schneider, H. *J. Biol. Chem.* **1986**, *261*, 10624–10631.
8. Cotter, R. J.; Honovich, J.; Qureshi, N.; Takayama, K. *Biomed. Environ. Mass Spectrom.* **1987**, *14*, 591–598.
9. Qureshi, N.; Takayama, K.; Mascagni, P.; Honovich, J.; Wong, R.; Cotter, R. J. *J. Biol. Chem.* **1988**, *263*, 11917–11976.
10. Wang, R.; Chen, L.; Cotter, R. J.; Qureshi, N.; Takayama, K. *J. Microbiol. Methods* **1992**, *15*, 151–166.
11. Qureshi, N.; Takayama, K.; Ribi, E. *J. Biol. Chem.* **1982**, *257*, 11808–11815.
12. Takayama, K.; Qureshi, N. In *Bacterial Endotoxic Lipopolysaccharides*; Morrison, D. C.; Ryan, J. L., Eds.; Molecular Biochemistry and Cellular Biology; CRC Press, Inc.: Boca Raton, FL, 1992, Vol. 1; pp. 43–66.
13. Seydel, U.; Lindner, B.; Wollenweber, H. -W.; Rietschel, E. T. *Eur. J. Biochem.* **1984**, *145*, 505–509.
14. Seydel, U.; Lindner, B.; Zähringer, U.; Rietschel, E. T.; Kusumoto, S.; Shiba, T. *Biomed. Mass Spectrom.* **1984**, *11*, 132–141.
15. Tabet, J. -C.; Cotter, R. J. *Anal. Chem.* 1984, *56*, 1662–1667.
16. Spengler, B.; Dolce, J. W.; Cotter, R. J. *Anal. Chem.* **1990**, *62*, 1731–1737.
17. Qureshi, N.; Honovich, J. P.; Hara, H.; Cotter, R. J.; Takayama, K. *J. Biol. Chem.* **1988**, *263*, 5502–5504.
18. Qureshi, N.; Takayama, K.; Meyer, K. C.; Kirkland, T. N.; Bush, C. A.; Chen, L.; Wang, R.; Cotter, R. J. *J. Biol. Chem.* **1991**, *266*, 6532–6538.
19. Westphal, O.; Jann, K. *Methods Carbohydr. Chem.* **1965**, *5*, 83–91.
20. Galanos, C.; Lüderitz, O.; Westphal, O. *Eur. J. Biochem.* **1969**, *9*, 245–249.
21. Raetz, C. R. H.; Purcell, S.; Meyer, M. V.; Qureshi, N.; Takayama, K. *J. Biol. Chem.* **1985**, *260*, 16080–16088.
22. Qureshi, N.; Mascagni, P.; Ribi, E.; Takayama, K. *J. Biol. Chem.* **1985**, *260*, 5271–5278.

RECEIVED May 14, 1993

Chapter 13

Lipooligosaccharides in Pathogenic *Haemophilus* and *Neisseria* Species

Mass Spectrometric Techniques for Identification and Characterization

Bradford W. Gibson, Nancy J. Phillips, Constance M. John[1], and William Melaugh

Department of Pharmaceutical Chemistry, School of Pharmacy, University of California, San Francisco, CA 94143-0446

Mass spectrometric methods have been developed for the analysis of lipooligosaccharides (LOS) from pathogenic *Neisseria* and *Haemophilus* species. Since LOS from these species are much smaller in size (M_r ~3000-7000) than the lipopolysaccharides (or LPS) of most other Gram-negative bacteria, a modified strategy is used to characterize these glycolipids. Currently, negative-ion electrospray ionization mass spectrometry (ESI-MS) is used to determine the overall heterogeneity and establish preliminary compositions of the LOS molecules produced from a single bacterial strain after *O*-deacylation with mild hydrazine. Subsequent analyses of the oligosaccharide regions produced by mild acid treatment of the LOS are then carried out by a combination of liquid secondary ion mass spectrometry (LSIMS), tandem mass spectrometry and conventional techniques such as methylation analysis and NMR. Taken together, these data provide critical information on the diverse population of LOS in which the primary structural variation resides in the carbohydrate portion.

Gram-negative bacteria belonging to *Neisseria* and *Haemophilus* species constitute a diverse group of organisms. Some species, such as *Neisseria lactamica* are commensal and non-pathogenic, and colonize the mouth and throat without apparent damage to the human host. Other strains such as *Neisseria gonorrhoeae* and *Haemophilus ducreyi* are highly pathogenic, causing the sexually transmitted diseases gonorrhea and chancroid (*1*) (a genital ulcer disease). In women, gonococcal infection can remain asymptomatic for some time, yet ultimately cause pelvic inflammatory disease with serious long-term health consequences (*2*). *Haemophilus influenzae* is primarily a respiratory pathogen, colonizing both pulmonary and ear mucosal surfaces. Type b *H. influenzae* infections are one of the major causes of childhood meningitis, but *H. influenzae* is also capable of causing pneumonia, bacteremia, and conjunctivitis (*3*).

Considerable work has been carried out to determine the virulence factors associated with the pathogenic *Neisseria* and *Haemophilus* species. Our own studies have concentrated on the structure and biological role(s) of surface

[1]Current address: Charles Evans and Associates, 310 Chesapeake Drive, Redwood City, CA 94063

lipooligosaccharides (LOS), one of the major outer membrane components of these bacteria (4). LOS differ from lipopolysaccharides (or LPS) produced by most Gram-negative bacteria in that they are considerably smaller in size (M_r 3000-7000), and do not form large repeating glycan structures that characterize LPS (5). A generalized structure of LOS is presented in Figure 1 for *Neisseria* and *Haemophilus* strains. Key features of these LOS are the relatively conserved lipid A (6,7) and core regions that serve to anchor a highly complex and variable oligosaccharide branch region. This branch region has been found to extend off either one or two of the core heptoses and, at least in some noted cases, contain terminal structures similar to glycans present in human glycosphingolipids (8-12). These terminal glycans are also implicated in phase variation (13), and may be important to the pathology of these organisms.

In this report, details of our mass spectrometric strategy that we are currently using to characterize the LOS from *Haemophilus ducreyi* and *H. influenzae*, and *Neisseria gonorrhoeae* and *N. meningitidis* are presented. Emphasis is given to the identification of the structural features contained in these LOS molecules that appear to be characteristic or unique to the particular species. Although we have recently begun characterization of the lipid A moieties for several of these bacteria, discussion in this report will focus only on the mass spectrometric methods used in the analyses of the variable glycan region. In principle, these methods could be equally well applied to other important pathogenic bacteria such as *Chlamydia* and *Bordetella pertussis*, which also make LOS rather than LPS as their primary surface glycolipids.

Methods

Preparation and Isolation of LOS. LOS from *H. ducreyi*, *N. gonorrhoeae* and *N. meningitidis* were isolated using a modified phenol water extraction method of Westphal and Jahn (14). The LOS from *Haemophilus influenzae* strains were prepared by the procedure of Darveau and Hancock (15).

Preparation and Purification of Oligosaccharide Fraction. LOS (2-20 mg) were hydrolyzed in 1% acetic acid (2 mg LOS/ml) for 2 h at 100°C. The hydrolysates were centrifuged at 5000 x g for 20 min at 4°C, and the supernatant containing the soluble oligosaccharide fraction was removed. The precipitates (containing primarily lipid A) were washed several times with 1-3 ml of H_2O and centrifuged again (5000 x g, 20 min, 4°C). The supernatant and washings were pooled and lyophilized. The lyophilized oligosaccharide fractions were dissolved in 0.5-1 ml of H_2O and centrifuge-filtered using Microfilterfuge tubes (Rainin Corp., MA).

Preparation and Analysis of *O*-deacylated LOS. LOS were first *O*-deacylated under mild hydrazine conditions (7). In most cases, 0.5-2 mg of LOS were incubated with 100-200 μl of anhydrous hydrazine for 20 min at 37°C. The samples were cooled to -20°C and 0.5-1 ml of chilled acetone was added dropwise to precipitate the *O*-deacylated LOS. The sample vials were then centrifuged at 12,000 x g for 20 min at 5°C, the supernatant removed, and the pellet washed again with cold acetone and centrifuged. The precipitated *O*-deacylated LOS was taken up in 0.5 ml of water and lyophilized.

A VG/Fisons Bio-Q mass spectrometer with an electrospray ion source operating in the negative-ion mode was used to mass analyze the *O*-deacylated LOS (16). The LOS samples were first dissolved in H_2O and 3 μl were injected via a Rheodyne injector into a constant stream of H_2O/CH_3CN (1/1, v:v) containing either 1% acetic acid or 0.1% triethylamine. Mass calibration was carried out with an external horse heart myoglobin reference using the supplied VG/Fison Bio-Q software.

Separation and Derivatization of Oligosaccharides. To separate oligosaccharide mixtures obtained after mild acid hydrolysis of LOS, several strategies were employed. In almost all cases, size-exclusion chromatography was first used to desalt the samples and obtain a partial purification of the glycan mixtures. In some cases, fractions of the oligosaccharides were derivatized with alkylphenylhydrazines (*17*) as shown in the reaction scheme detailed in Figure 2, and further separated by reverse phase HPLC. Briefly, oligosaccharides were derivatized with a hydrazine compound (preferably the butylphenylhydrazine) at an approx. molar excess over oligosaccharide of 2 to 1 for 30 min at 80°C. The reaction mixture was cooled, dried and then re-dissolved in water containing 0.05% trifluoroacetic acid for subsequent HPLC separation. A typical HPLC gradient consisted of a linear gradient of acetonitrile containing 0.05% trifluoroacetic acid (solvent B) added to 0.05% trifluoroacetic acid/H$_2$O (solvent A) from 0% - 50% B in 50 min. Alternatively, desalted oligosaccharide fractions were separated by high pH anion exchange chromatography (HPAEC) using a Dionex carbohydrate separation system equipped with a pulsed amperometric detector. This latter technique is not described in this report, but has been described in detail elsewhere (*18*).

LSIMS and Tandem Mass Spectrometry of Oligosaccharides and Oligosaccharide Derivatives. Both underivatized and derivatized oligosaccharides were directly analyzed by liquid secondary ion mass spectrometry (LSIMS). The oligosaccharides and their corresponding butylphenylhydrazine derivatives were dissolved in water. Samples were then analyzed on a Kratos MS 50S mass spectrometer retrofitted with a cesium ion source (*19*) and operating at a resolution of 1500-2000 (m/Δm, 10% valley). A primary ion beam of 10 keV was used to ionize the samples and secondary ions were accelerated at 6 kV. Scans were acquired at 300 s/decade and recorded on a Gould electrostatic recorder. Ultramark 1206 was used for manual calibration to an accuracy better than ± 0.2 Da.

Tandem mass spectra (MS/MS) were obtained on a four-sector Kratos Concept II HH mass spectrometer fitted with an optically-coupled 4% diode array detector on MS-II as previously described (*20*). A cesium ion beam energy of 18 keV produced molecular ions which were selected in MS-I and passed to the helium collision cell floated at a potential of 2 kV where the gas pressure was adjusted to attenuate the parent ion beam by two-thirds of its initial value. The product ions were detected and analyzed in MS-II with successive 4% frames and a constant B/E ratio. Fragment ion assignments for all carbohydrate mass spectra are made based on the proposed nomenclature of Domon and Costello (*21*).

Results

Molecular Weight Analysis of LOS. In our current strategy for the characterization and analysis of *Neisseria* and *Haemophilus* glycolipids, we use electrospray ionization mass spectrometry (ESI-MS) (*22*) to provide an overview of the glycoforms that are present in these LOS preparations (*16*). The molecular weight data in the ESI-MS spectra give us a clear indication of the precise sizes and, in many cases, monosaccharide compositions, of these LOS. These data correlate well with the SDS-PAGE analysis (*23*) and monoclonal antibody binding studies that are often available, and address precise structure-function relationships at an early stage of the analysis. By providing a good estimation, therefore, of the sizes and heterogeneity of these LOS, we are in a much better position to design fractionation and purification strategies for the subsequent oligosaccharide analyses that are more tailored to the individual LOS preparation.

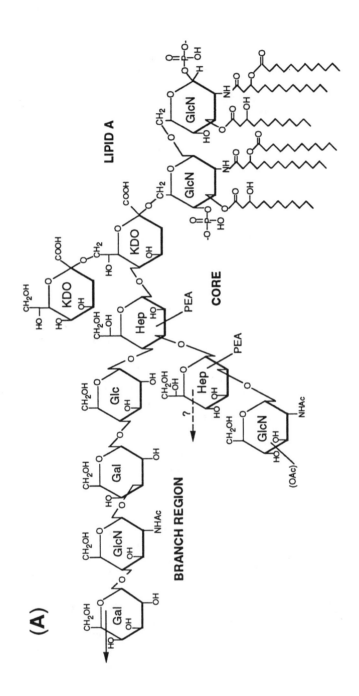

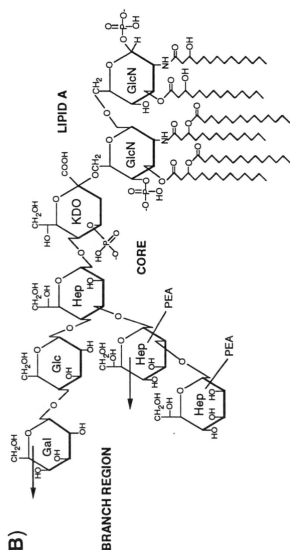

Figure 1. Generalized models for the structure of the LOS from (A) *Neisseria* and (B) *Haemophilus* species. The neisserial LOS is shown containing a terminal lacto-*N*-neotetraose as identified in many strains, including *N. gonorrhoeae* F62 and 1291. Phosphorylation heterogeneity of the core heptose region is evident by the presence of one or more phosphoethanolamine (PEA) moieties. The *Haemophilus* LOS is based on that determined for *H. influenzae* strain 2019, and the arrows indicate some LOS contain extended branches off the first and second core heptoses. In both *H. influenzae* and *H. ducreyi*, the KDO is phosphorylated on the 4-position. In *H. ducreyi*, a fourth heptose has been found on the branch region, and has been tentatively assigned as a D-*glycero*-D-*manno*-heptose. *Neisseria* and *Haemophilus* LOS are also sialylated most often, if not exclusively, on a terminal lactosamine (Galβ1→4GlcNAc) acceptor.

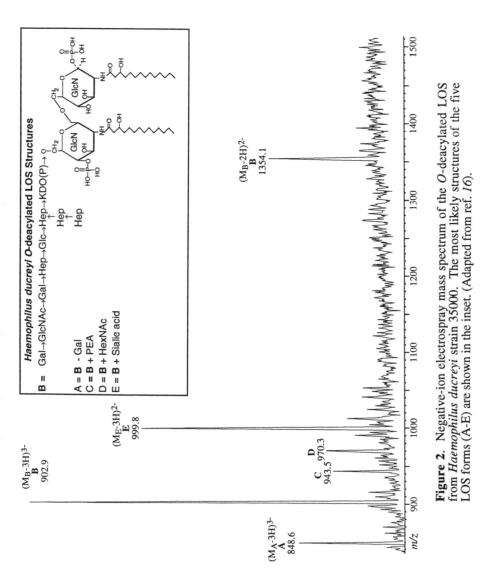

Figure 2. Negative-ion electrospray mass spectrum of the *O*-deacylated LOS from *Haemophilus ducreyi* strain 35000. The most likely structures of the five LOS forms (A–E) are shown in the inset. (Adapted from ref. *16*).

To date, we have not been able to analyze unmodified or intact LOS using either LSIMS or ESI-MS. This is likely due to the heterogeneity of the LOS preparations, high masses (at least for LSIMS), and low solubility in solvents compatible with mass spectrometry. Fortunately, a simple and relatively mild procedure of hydrazine treatment will form *O*-deacylated LOS species which are now water soluble and quite amenable to mass spectrometric analyses (*18*). Mild hydrazine treatment of small amounts of material quantitatively removes all *O*-linked acyl moieties, producing *N*-linked diacyl lipid A species from their native hexaacyl forms.

In Figure 2, the ESI-MS spectrum of *O*-deacylated LOS from *H. ducreyi* strain 35000 is shown. This spectrum reflects the structural heterogeneity typically found in most LOS preparations of *Haemophilus* and *Neisseria* (*16*). For example, it is apparent from the number of peaks that there are several components in this LOS preparation. The base peak at *m/z* 902.9 can be readily assigned as the triply charged deprotonated molecular ion, $(M-3H)^{3-}$, with the only other member of this molecular ion series at *m/z* 1354.1, $(M-2H)^{2-}$. Under negative-ion conditions, most *O*-deacylated LOS usually give only one or two charge states that are largely determined by the number of phosphates. Since the lipid A of *Haemophilus* contains two phosphates and the KDO one phosphate, the typical charge state is $z = -3$. In this spectrum, mass differences from the base peak to nearby peaks can be used to readily assign the remaining peaks as triply charged ions corresponding to the loss of hexose (*m/z* 848.6), or the addition of PEA (*m/z* 943.5), HexNAc (*m/z* 970.3) or sialic acid (*m/z* 999.8).

The presence of sialic acid can be confirmed in these LOS mixtures using ESI-MS in combination with neuraminidase treatment (*24*). For example, treatment of this same *H. ducreyi* *O*-deacylated LOS preparation with neuraminidase, followed by microdialysis and re-analysis by ESI-MS results in the loss of this presumed sialic acid-containing species at *m/z* 999.8, with a concurrent increase in the abundance of the asialo-LOS counterpart at *m/z* 902.9 (unpublished data, B. Gibson). This mass spectrometric method represents a considerable improvement over the conventional analytical technique where sialic acid content is measured on a weight basis i.e., μg sialic acid/mg LOS) (*25*). In the ESI-MS method, sialic acid can be assigned to specific LOS species, as well as a relative mole percent of sialylated LOS based on the relative peak abundances to the asialo LOS components.

General Strategy for Structure Determination of the Oligosaccharides. While electrospray ionization mass spectrometric data provides considerable insight into the overall structures and possible saccharide compositions present in a particular LOS preparation, further information is needed at the oligosaccharide level to address the precise nature of the variable and heterogeneous glycan moieties. Data of this type can be provided by mass spectrometric analysis in several ways. Molecular masses of the oligosaccharides can be obtained to an accuracy of better than ± 0.2 Da by direct LSIMS analysis, clearly sufficient to work out the possible sugar compositions. However, to delineate in more detail the oligosaccharide structures requires either further purification of the individual oligosaccharides to homogeneity and/or tandem mass spectrometry of these oligosaccharides. In the subsequent sections, data are shown that represent what can generally be obtained from analysis of oligosaccharides, either before or after derivatization, by LSIMS under conventional or tandem high energy collision induced dissociation (CID) conditions.

LSIMS of free Oligosaccharides. In most cases, the oligosaccharide pool generated by mild acid hydrolysis of LOS is partially separated (and desalted) by size exclusion chromatography (*18*). In almost all cases, however, fractions obtained after this type of chromatographic purification are invariably mixtures. This is due in many cases to heterogeneity in phosphorylation or O-acetylation states, but is also due to variable sugar compositions.

Shown below in Figure 3 is the negative-ion LSIMS spectrum of a separated fraction from the oligosaccharide pool of the LOS from N. gonorrhoeae strain F62. It is quite clear from this spectrum that there are several oligosaccharides present in this single fraction. Each oligosaccharide exists with and without an O-acetyl group as indicated by the mass difference of 42 Da that separates the two main peaks in each peak cluster. For example, (M-H)⁻ ions are seen at m/z 1759 and 1801, and m/z 1962 and 2004. In addition to these ions, their corresponding sodiated counterparts, (M-2H+Na)⁻, are also apparent at m/z 1781 and 1823, and m/z 1984 and 2026, respectively. The sodiated oligosaccharide species are present at relatively high abundances despite efforts taken to desalt these samples. After HF treatment of these fractions, however, the sodiated peaks virtually disappear as the masses of the different oligosaccharides shift 246 Da lower, consistent with the presence of two phosphoethanolamine moieties (see Figure 4). This suggests that the persistence of the sodiated species in the non-HF samples is due to the preferred binding of sodium to either of these two phosphoethanolamine moieties. The mass difference of 203 Da between these two peaks also suggests that the larger components contain an additional HexNAc moiety. Based on these masses alone, preliminary compositions can be assigned that provide a working hypothesis for the continuing structural studies (see Table I). The assignments for two of these oligosaccharides have, in fact, been confirmed in a recently published NMR study of the two major oligosaccharides after HF treatment (26). The other assignments have also been confirmed by subsequent composition and methylation analyses (27).

Table I. N. *gonorrhoeae* Strain F62 Oligosaccharide Analysis by LSIMS

(M-H)⁻	Proposed Compositions
1759	Hex₃ HexNAc₂ Hep₂ PEA₂ KDO
1801	Hex₃ HexNAc₂ (Ac) Hep₂ PEA₂ KDO
1962	Hex₃ HexNAc₃ Hep₂ PEA₂ KDO
2004	Hex₃ HexNAc₃ (Ac)Hep₂ PEA₂ KDO
2124	Hex₄ HexNAc₃ Hep₂ PEA₂ KDO
2166	Hex₄ HexNAc₃ (Ac) Hep₂ PEA₂ KDO

The complexity of the LSIMS spectra of gonococcal oligosaccharides is not always as large as shown in Figure 3, but is typical of wild type oligosaccharide pools from the LOS from both N. gonorrhoeae and N. meningitidis. Haemophilus oligosaccharides have not yet been found to contain O-acetyl groups, and generally show only one major peak for each oligosaccharide component. If one treats neiserrial oligosaccharides with HF to remove phosphate esters and O-acetyl groups, a much less complex molecular ion population is observed (see Figure 4). Fragment ions can also be observed when the oligosaccharides are present in relatively pure form, as evident from the presence of two sets of reducing terminal Y- and Z-type fragments in the mass spectrum of this HF-treated oligosaccharide. These fragments establish at least part of the two non-reducing terminal branch regions as either HexNAc→ and Hex→HexNAc→ or, alternatively, Hex and HexNAc→Hex. Methylation analysis was used to distinguish between these two possibilities, which clearly showed that both Gal and GalNAc are present as non-reducing terminal sugars. When immuochemical data is considered as well, the two branches can be assigned as Galβ1→4GlcNAc and GalNAc1→. (See Figure 5.)

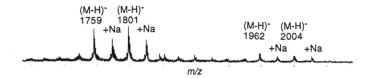

Figure 3. Partial negative-ion LSIMS spectrum of a late eluting fraction from a Bio-Gel P-4 separation of the underivatized oligosaccharides from *N. gonorrhoeae* strain F62. (M-H)⁻ ions at m/z 2124 and 2166 listed in Table I are not present in this spectrum and were observed only in the early eluting fractions (data not shown).

LSIMS of Hydrazone Oligosaccharides. Although one can sometimes obtain sequence information from the underivatized oligosaccharide spectra, it is generally the case that derivatives of one type or another yield more extensive sequence information than the free oligosaccharide itself. Permethylation of these glycans can provide more structurally useful fragments, but is complicated by side reactions that involve the partial destruction of the reducing terminal KDO (*28,29*), as well as requiring a relatively large amount of material for the chemistry workup. Our efforts have therefore been concentrated on the preparation of derivatives that involve only a single chemical step, and at the same time produce an oligosaccharide derivative amenable to further chromatographic purification. The best derivatives we have found to date for these types of glycans are the alkylphenylhydrazines or their benzoate counterparts (*17*). The chemistry is straightforward and can be carried out at very low levels. The resulting products can be separated by reverse phase HPLC prior to mass spectrometric analysis due to their increased hydrophobicity and through the presence of a UV absorbing chromophore (see Figure 6).

A recent study on the oligosaccharides from a strain of *H. ducreyi* (*12*) illustrates some of the advantages of this technique as well as some potential undesirable complications. In Figure 7, the spectra of the major oligosaccharide from *H. ducreyi* strain 35000 is shown as its corresponding butylphenylhydrazine derivative. It is quite clear that the extent of fragmentation is considerable, and is much improved over underivatized species. Fragment ions are seen originating from the cleavage of almost all the glycosidic bonds, with charge retention exclusively on the reducing terminus. Interpretation of the spectra is relatively straightforward. Nonetheless, cleavage between the two inner-core heptoses is not observed, and the branch glycan structure cannot be linked to a specific core heptose.

Surprisingly, the HPLC chromatogram of the hydrazone-oligosaccharides from this *H. ducreyi* strain (see inset, Figure 7) was considerably more heterogeneous than expected from our previous analysis of the underivatized oligosaccharide pool by LSIMS. Furthermore, each peak gave the *same* molecular ion as its base peak, i.e. (M-H)⁻ at *m/z* 1822. NMR analysis confirmed the presence of various anhydro-KDO forms in this oligosaccharide that were apparently formed by β-elimination of 4-linked phosphate during acetic acid hydrolysis of the LOS (see Figure 8). *H. influenzae* LOS also contain a phosphorylated KDO which undergoes an identical reaction during the acetic acid hydrolysis step (*10*). Due to this problem, we currently limit our use of the hydrazine derivatives to oligosaccharides derived from *Neisseria* species, or to *Haemophilus* LOS that have first undergone HF treatment. Removal of this 4-linked phosphate by HF prior to hydrolysis precludes the

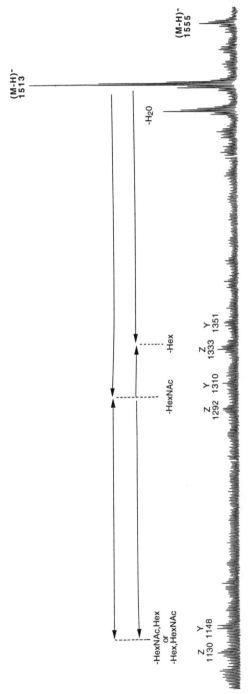

Figure 4. Partial negative-ion LSIMS spectrum of a dephosphorylated oligosaccharide fraction originating from the same sample shown in Figure 3. This dephosphorylated oligosaccharide sample has also undergone size exclusion separation after HF-treatment and now contains only the dephosphorylated analogs of the major (M-H)⁻ species at m/z 1759 and 1801.

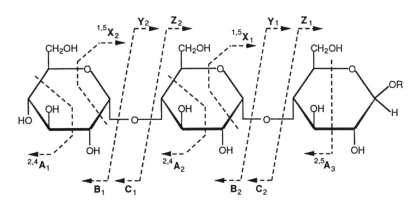

Figure 5. Fragmentation types and nomenclature for carbohydrates. (Adapted from ref. *21*).

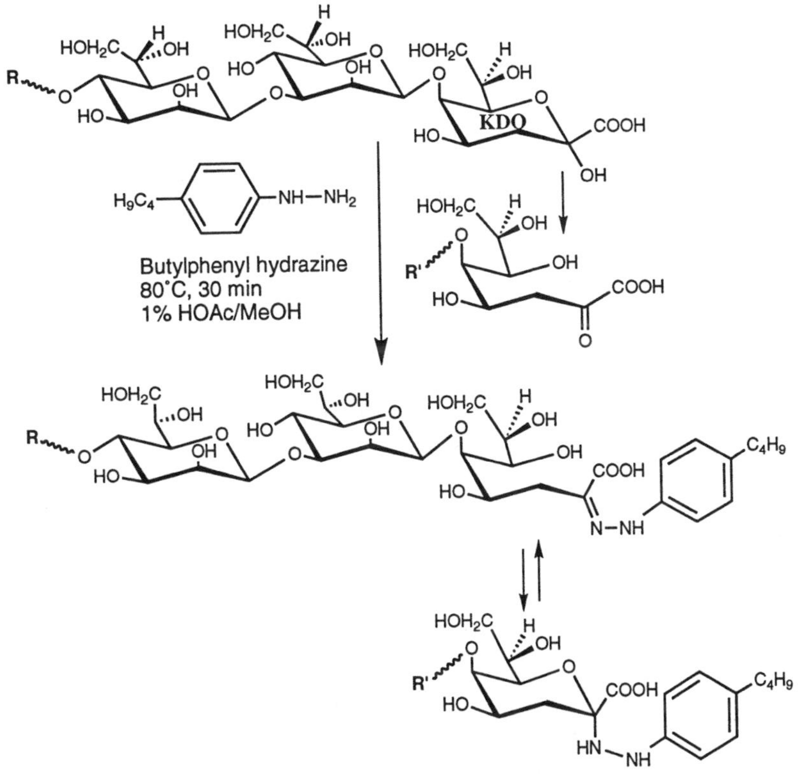

Figure 6. Derivatization scheme for the formation of butylphenylhydrazine oligosaccharide derivatives. In *Haemophilus*, β-elimination of a 4-linked phosphate during acetic acid hydrolysis results in the formation of anhydro-KDO species that can form multiple reaction products upon derivatization with butylphenylhydrazine.

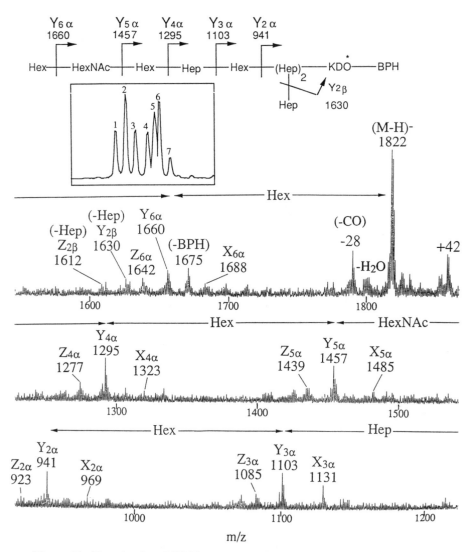

Figure 7. Negative-ion LSIMS spectrum of butylphenylhydrazine derivative (peak 2) of the major oligosaccharide from *H. ducreyi* strain 35000 after (inset) HPLC chromatography. (Adapted from ref. *12*)

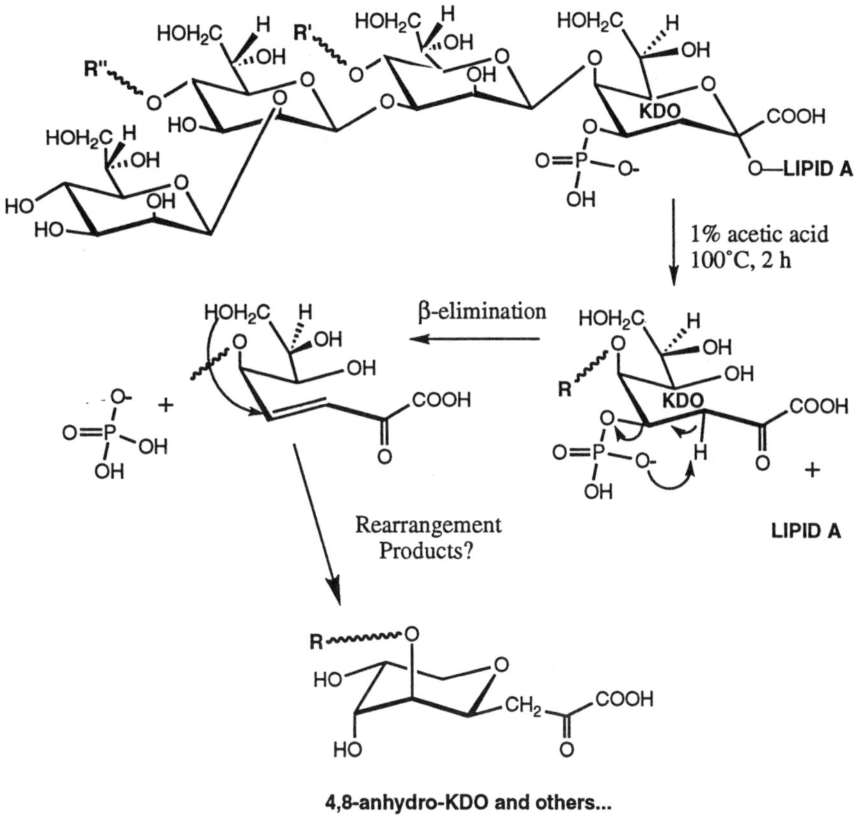

Figure 8. β-elimination of phosphate during the acetic acid hydrolysis of *Haemophilus* LOS and one of the possible anhydro-KDOs formed after rearrangement of the α,β-unsaturated ketone intermediate.

formation of this elimination product and reduces the number of hydrazino forms seen during HPLC purification.

Tandem MS of Free and Derivatized Oligosaccharides. As mentioned in the previous sections, fragment ion data contained in the LSIMS spectra of oligosaccharides or their corresponding oligosaccharide derivatives are not generally sufficient to assign an *overall* structure. In particular, the fragment ions necessary to assign the different branches to their respective heptoses are often absent. Under high-energy CID conditions, however, even the resistant core heptose region produces extensive fragments, thereby allowing for the determination of an overall structure. Formation of internal fragments via cleavage processes at more than one remote site can sometimes appear in these CID spectra, but there abundances are generally low compared to the typically observed single site cleavage processes.

In Figure 9, a typical tandem CID spectrum of an underivatized oligosaccharide is shown for an octasaccharide derived from the LOS of *H. influenzae* strain A2. Under CID conditions, fragmentation is seen at every glycosidic bond, as well as ring cleavage reactions (cleavage type **A**) that are more prevalent under high energy CID conditions. When phosphate esters are present the observed ion fragments in the negative-ion CID spectra are generally biased towards those fragments containing the phosphate moiety. Presumably this is due to the much preferred charge stabilization of the phosphate anion. An important ring fragment ion is also seen at m/z 506 and can be attributed to the major α-branch ($^{2,4}A_{3\alpha}$). This interpretation is consistent with other data suggesting that the phosphoethanolamine is substituted at the 4-position of the middle core heptose.

In addition to the major **A**, **Y** and **B**-type fragments observed in the tandem spectrum of this *H. influenzae* A2 oligosaccharide, some smaller fragment ions were observed at m/z 622 ant 654 that are not consistent with any single cleavage process. We have tentatively assigned these product ions as originating from the Glc→Glc→Hep(PEA) core region, involving cleavage at *both* the C1 and C2 glycosidic bonds of the middle heptose (i.e., an internal fragment). The mass difference between these two ions is consistent with the loss of O_2 which might originate from a peroxide-containing intermediate formed from the neighboring heptose oxygens previously linked to the other two heptoses. Further studies will be needed to establish this possibility.

One can also obtain good tandem CID spectra from derivatized oligosaccharides, as shown in Figure 10 for a pyocin-mutant of *N. gonorrhoeae* 1291_c (*11*). Since this oligosaccharide had been previously treated with HF to remove phosphoethanolamine moieties, the observed fragment ions are distributed more randomly than those shown for the *H. influenzae* oligosaccharide in Figure 9. Fragment ions are seen resulting from the cleavage of every glycosidic bond. Typical of CID spectra is the presence of an extensive set of ring cleavage ions with charge retention at both the reducing (*X*-type) and non-reducing terminus (*A*-type). Together with other information obtained from methylation and composition analyses and immunochemical studies, the structures of these glycan can be assembled in a relatively rapid fashion.

Summary

Mass spectrometric analysis of relatively intact LOS forms or their separated glycan (or lipid A) moieties can be used to structurally characterize these diverse and complex glycolipids. *Neisseria* and *Haemophilus* species contain significant differences in both their lipid A and glycan structures. These structural differences appear to be unique to these particular bacteria, and through further correlation of LOS structures with biological and immunochemical data can provide insight into their roles in the disease states they cause.

It has already been established that many *Neisseria* and *Haemophilus* species and strains contain glycans at their exposed non-reducing termini that are similar if not identical to glycans present on human glycosphingolipids. This host mimicry may underlie the ability of these organisms to evade the immune system. Alternatively, these glycans may provide a means to adhere to host cells, or be involved in host cell invasion mechanisms. A comparative study of the LOS from several different *Haemophilus* and *Neisseria* species, along with biological and immunological data, should enable us to truly understand the underlying molecular basis of LOS roles in the pathology of these organisms.

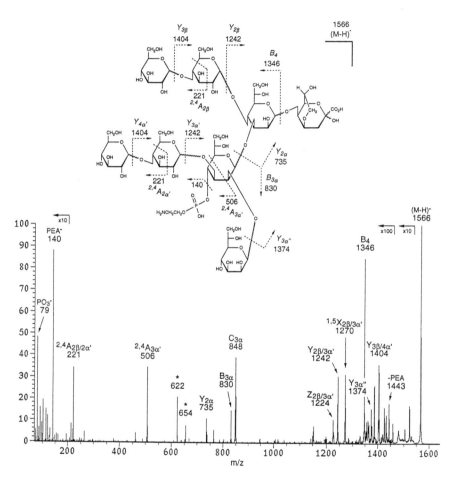

Figure 9. Tandem CID spectrum of underivatized oligosaccharide from *H. influenzae* strain A2. Note the presence of *m/z* 140 for deprotonated phosphoethanolamine and its secondary fragment at *m/z* 79 (PO₃⁻). (Reproduced with permission from ref. *24*. Copyright 1993 American Chemical Society).

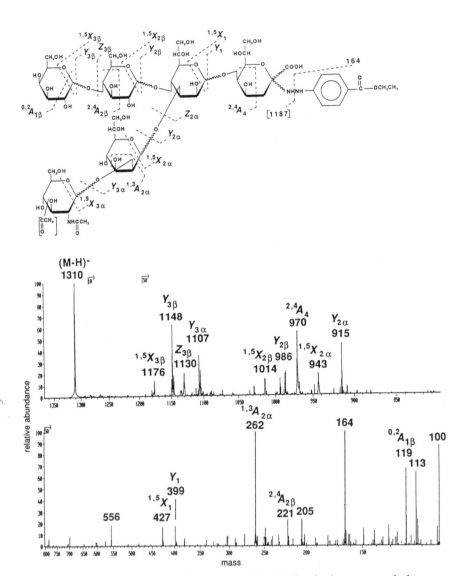

Figure 10. Tandem CID spectrum of the hydrazinobenzoate ethyl ester derivative of the oligosaccharide isolated from the LOS of *N. gonorrhoeae* pyocin-mutant strain 1291$_C$. (Reproduced with permission from ref. *11*. Copyright 1991 The Journal of Biological Chemistry).

Acknowledgments

We would like to acknowledge the contributions of our colleagues M. Apicella and A. Campagnari (SUNY at Buffalo), J. McLeod Griffiss, R. Yamasaki, J. Kim and R. Mandrell (Veterans Administration, UCSF), and grant support from the National Institutes of Health for work on *Neisseria gonorrheae* (AI 21620 and AI 18384), *Neisseria meningitidis*, and *Haemophilus influenzae*. (AI 24616). Structural studies of *H. ducreyi* have been supported by the Universitywide AIDS Research Program (R 91SF243). We would also like to acknowledge the UCSF mass spectrometry facility which is supported by grants from the NIH (RR 01614) and NSF Biological Instrumentation Program (DIR 8700766).

Literature Cited

1. Johnson, A. P.; Abeck, D.; Davies, H. A. *J. Infect.* **1988**, *17*, 99-106.
2. Westrom, L. *Am. J. Obstec. Gynecol.* **1980**, *138*, 880-892.
3. Moxon, E. R. In *Principles and Practice of Infectious Diseases*; 3rd Edition; G. L. Mandell, J. R. G. Douglas and J. E. Bennett, Eds.; Churchill Livingstone Inc.: New York, 1990; pp 1722-1729.
4. Griffiss, J. M.; Schneider, H.; Mandrell, R. E.; Yamasaki, R.; Jarvis, G. A.; Kim, J. J.; Gibson, B. W.; Hamadeh, R.; Apicella, M. A. *Rev. Infect. Dis.* **1988**, *10*, s287-s295.
5. Wilkinson, S. G. In *Surface Carbohydrates of the Prokaryotic Cell*; I. Sunderland, Ed.; Academic Press: New York, 1977; pp 97-175.
6. Takayama, K.; Qureshi, N.; Hyver, K.; Honovich, J.; Cotter, R. J.; Mascagni, P.; Schneider, H. *J. Biol. Chem.* **1986**, *261*, 10624-10631.
7. Helander, I. M.; Lindner, B.; Brade, H.; Altmann, K.; Lindberg, A. A.; Rietschel, E. T.; Zähringer, U. *Eur. J. Biochem.* **1988**, *177*, 483-492.
8. Mandrell, R. E.; McLaughlin, R.; Abu Kwaik, Y.; Lesse, A.; Yamasaki, R.; Gibson, B.; Spinola, S. M.; Apicella, M. A. *Infect. Immun.* **1992**, *60*, 1322-1328.
9. Mandrell, R. E.; Griffiss, J. M.; Macher, B. A. *J. Exp. Med.* **1988**, *168*, 107-126.
10. Phillips, N. J.; Apicella, M. A.; Griffiss, J. M.; Gibson, B. W. *Biochemistry* **1992**, *31*, 4515-4526.
11. John, C. M.; Griffiss, J. M.; Apicella, M. A.; Mandrell, R. E.; Gibson, B. W. *J. Biol. Chem.* **1991**, *266*, 19303-19311.
12. Melaugh, W.; Phillips, N. J.; Campagnari, A. A.; Karalus, R.; Gibson, B. W. *J. Biol. Chem.* **1992**, *267*, 13434-13439.
13. Maskell, D. J.; Szabo, M.; Butler, P. D.; Williams, A. E.; Moxon, E. R. *Res. Microbiol.* **1991**, *142*, 719-724.
14. Westphal, O.; Jahn, K. *Methods Carbohydr. Chem.* **1954**, *5*, 83-91.
15. Darveau, R. P.; Hancock, R. E. W. *J. Bacteriol.* **1983**, *155*, 831-838.
16. Gibson, B. W.; Melaugh, W.; Phillips, N. J.; Apicella, M. A.; Campagnari, A. A.; Griffiss, J. M. *J. Bacteriol.* **1993**, in press.
17. John, C. M.; Gibson, B. W. *Anal. Biochem.* **1990**, *187*, 8523-8527.
18. Phillips, N. J.; John, C. M.; Reinders, L. G.; Gibson, B. W.; Apicella, M. A.; Griffiss, J. M. *Biomed. Environ. Mass Spectrom.* **1990**, *19*, 731-745.
19. Falick, A. M.; Wang, G. H.; Walls, F. C. *Anal. Chem.* **1986**, *58*, 1308-1311.
20. Walls, F. C.; Baldwin, M. A.; Falick, A.; M.; Gibson, B. W.; Kaur, S.; Maltby, D. A.; Gillece-Castro, B. L.; Medzihradszky, K. F.; Evans, S.; Burlingame, A. L. In *Biological Mass Spectrometry*; A. L. Burlingame and J. A. McCloskey, Eds.; Elsevier: Amsterdam, 1990; pp 197-216.
21. Domon, B.; Costello, C. E. *Glycoconjugate J.* **1988**, *5*, 397-409.

22. Fenn, J. B.; Mann, M.; Meng, C. K.; Wong, S. F.; Whitehouse, C. M. *Science* **1989**, *246*, 64-71.
23. Hitchkock, P. J. *Infect. Immun.* **1984**, *46*, 202-212.
24. Phillips, N. J.; Apicella, M. A.; Griffiss, J. M.; Gibson, B. W. *Biochemistry* **1993**, *32*, 2003-2012.
25. Mandrell, R. E.; Kim, J. J.; John, C. M.; Gibson, B. W.; Sugai, J. V.; Apicella, M. A.; Griffiss, J. M.; Yamasaki, R. *J. Bacteriol.* **1991**, *173*, 2823-2832.
26. Yamasaki, R.; Bacon, B. E.; Nasholds, W.; Schneider, H.; Griffiss, J. M. *Biochemistry* **1991**, *30*, 10566-10575.
27. John, C. M. Ph.D. Thesis, University of California, San Francisco, **1990**.
28. Gibson, B. W.; Webb, J. W.; Yamasaki, R.; Fisher, S. J.; Burlingame, A. L.; Mandrell, R. E.; Schneider, H.; Griffiss, J. M. *Proc. Natl. Acad. Sci. USA* **1989**, *86*, 17-21.
29. Dell, A.; Azadi, P.; Tiller, P.; Thomas-Oates, J.; Jennings, H. J.; Beurret, M.; Michon, F. *Carbohydr. Res.* **1990**, *200*, 59-76.

RECEIVED April 16, 1993

Chapter 14

The Glycolipids of Mycobacteria

G. S. Besra and P. J. Brennan

Department of Microbiology, Colorado State University, Fort Collins, CO 80523

Research over the past 15 years has shown that members of the Mycobacterium genus express on their surfaces large quantities of glycolipids of highly unusual structures. Members of the Mycobacterium avium complex, important opportunistic pathogens, are characterized by glycopeptidolipids, and serotypic variation in this complex is due to variations in the composition of the short oligosaccharide haptens of the glycopeptide lipid antigens. Other atypical mycobacteria (environmental; sometimes pathogenic) are marked by the lipooligosaccharides, i.e., oligosaccharides based on an acylated trehalose core. Another major group of glycolipids, found to best effect in Mycobacterium leprae, are the phenolic glycolipids based on a phenolphthiocerol. In this review, emphasis is placed on structure, especially glycosyl composition and sequence.

The resurgence of interest in mycobacteria and their constituents stems from the resurgence of tuberculosis itself (*1*) and from long-time fascination with leprosy. Similarly, the involvement of "atypical," nontuberculous environmental mycobacteria, such as the members of the *Mycobacterium avium* complex, as opportunistic pathogens in patients with underlying immune dysfunctions (*2*) has mitigated interest in the bacteria themselves and also in their peculiar immunopathogenesis.

Most, if not all, of the nontuberculous mycobacterial species are endowed with large quantities of a variety of glycolipid antigens with remarkable structural features (*3*). These glycolipids, and the more complex lipopolysaccharides that accompany them, have been implicated in diverse aspects of disease pathogenesis, such as interaction of bacterium and macrophage and persistence of bacteria within the intracellular environment (*4*). The variable oligosaccharide constituents of these glycolipid antigens are usually of sufficient

0097–6156/94/0541–0203$08.50/0

antigenicity as to evoke corresponding specific antibodies and thereby allow serodiagnosis of individual mycobacterioses (3), and, in the case of leprosy, a good means for specific serodiagnosis (5). Three major classes of glycolipids with the generic structures shown in Figure 1 have been identified, the trehalose-containing lipooligosaccharides (LOS) (6, 7), the glycopeptidolipids (GPL) (8), and the phenolic glycolipids (PGL) (3) (Figure 1).

The Characteristic Lipooligosaccharide Antigens of *Mycobacterium* spp. (Table I)

The acyltrehalose-containing lipooligosaccharides (LOS) are the predominant surface immunogens of a host of environmental mycobacteria and represent a rare principle in carbohydrate chemistry, glycosidically linked trehalose. The acyltrehalose residue invariably occupies the putative reducing terminus, whereas the non-reducing end is variable, being responsible for immunological specificity, and is characteristic of each individual species. First members of this class were reported by Saadat and Ballou (9) who called substances in *M. smegmatis* pyruvylated glycolipids since they contain pyruvate ketal units. Initially, Saadat and Ballou were seeking biosynthetic precursors of the 6-*O*-methylglucose lipopolysaccharide and had isolated from cell extracts of *M. smegmatis* several acidic methylglucose-containing glycolipids. The acidic property was found to be due to the presence of a pyruvate group, rather than the glycerate that characterizes the methylglucose-containing lipopolysaccharide. More recently, a pyruvylated LOS glycolipid was isolated from *Mycobacterium butyricum* (10) which differed solely by the extent of acylation of the oligosaccharide core as compared to the acidic oligosaccharide A derived from the pyruvylated glycolipid of Saadat and Ballou (9) (Table I).

Schaefer had reported that the surface antigens of *Mycobacterium kansasii* involved in specific agglutination were soluble in organic solvents, and their antigenic properties, unlike those from *M. avium*, were susceptible to mild alkali treatment (11). Hunter *et al.* (6) demonstrated that the serologically active lipids of *M. kansasii*, previously described by Schaefer, in fact, belonged to the new class of glycosylated acylated trehaloses. The core structure of eight of these glycolipids from *M. kansasii* (Figure 2) were revealed by a combination of acetolysis, partial acid cleavage, ^{1}H- and ^{13}C-NMR spectroscopy, chemical ionization and EI/MS of permethylated products (6, 7, 12). They are composed of variable residues of xylose, 3-*O*-methylrhamnose, fucose, and a novel *N*-acylamino sugar linked to a common tetraglucose core, distinguished by the presence of a glycosidically linked α,α'-trehalose substituent. The novel *N*-acyl amino sugar was later shown to be 4,6-dideoxy-2-*O*-CH$_3$-3-CH$_3$-4-(2'-methoxypropionamido)-α-L-mannohexopyranosyl (12), and when the terminal, non-reducing end *N*-acylkansoaminyl(1→3)fucosyl disaccharide appeared on the more polar glycolipids (LOS IV-VIII), antigenicity, and precise anti-*M. kansasii* antigenicity at that, was conferred on these molecules. Thus, this unit is the unique antigenic epitope of *M. kansasii*, and the amino sugar *N*-acylkanosamine is the single most characteristic feature of the organism. Three molecules of 2,4-

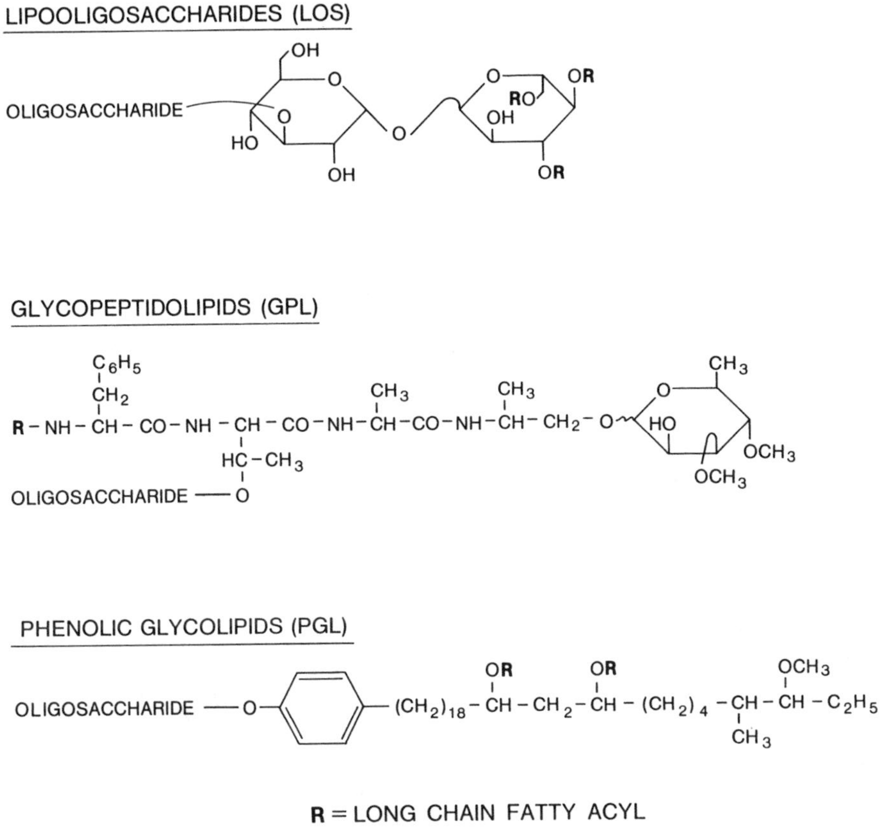

LIPOOLIGOSACCHARIDES (LOS)

GLYCOPEPTIDOLIPIDS (GPL)

PHENOLIC GLYCOLIPIDS (PGL)

R = LONG CHAIN FATTY ACYL

Figure 1. Generic structures of the three major classes of "new" glycolipids of *Mycobacterium* spp. In the case of the lipooligosaccharide class, the point of attachment of the oligosaccharide unit may also be the 4- or the 6-OH group of the acyltrehalose unit.

Table I   Structures of Major Trehalose - Containing Lipooligosaccharides of *Mycobacterium*

Species	Trivial Name	Structure of Oligosaccharide	Positions of Acyl Residues	Ref.
M. smegmatis	(Acidic oligosaccharide A)	4,6-(1′-carboxyethylidene)-3-*O*-Me-β-D-Glc*p*-(1→3)-4,6-(1′-carboxyethylidene)-β-D-Glc*p*-(1→4)-β-D-Glc*p*-(1→6)-α-D-Glc*p*-(1↔1)-α-D-Glc*p*	4 and 6-hydroxyls of terminal trehalose	9
M. smegmatis	(Acidic oligosaccharide B₁)	4,6-(1′-carboxyethylidene)-β-D-Glc*p*-(1→4)-β-D-Glc*p*-(1→6)-α-D-Glc*p*-(1↔1)-α-D-Glc*p*	ND[a]	
M. smegmatis	(Acidic oligosaccharide B₂)	4,6-(1′-carboxyethylidene)-3-*O*-Me-β-D-Glc*p*-(1→3)-β-D-Glc*p*-(1→6)-α-D-Glc*p*-(1↔1)-α-D-Glc*p*	ND	
M. kansasii	LOS I′	3-*O*-Me-α-L-Rha*p*-(1→3)-β-D-Glc*p*-(1→3)-β-D-Glc*p*-(1→4)-α-D-Glc*p*-(1↔1)-α-D-Glc*p*	ND	
M. kansasii	LOS I	β-D-Xyl*p*-(1→4)-3-*O*-Me-α-L-Rha*p*-(1→3)-β-D-Glc*p*-(1→3)-β-D-Glc*p*-(1↔1)-α-D-Glc*p*	ND	
M. kansasii	LOS II, III	(β-D-Xyl*p*)₂-(1→4)-3-*O*-Me-α-L-Rha*p*-(1→3)-β-D-Glc*p*-(1→3)-β-D-Glc*p*-(1↔1)-α-D-Glc*p*	ND	
M. kansasii	LOS IV, V, VI	KanNacyl-(1→3)-Fuc*p*-(1→4)-(β-D-Xyl*p*)₄-3-*O*-Me-α-L-Rha*p*-(1→3)-β-D-Glc*p*-(1→3)-β-D-Glc*p*-(1↔1)-α-D-Glc*p*	ND	
M. kansasii	LOS VII, VIII	KanNacyl-(1→3)-Fuc*p*-(1→4)-[β-L-Xyl*p*-(1→4)]₆-α-L-3-*O*-Me-Rha*p*-(1→3)- β-D-Glc*p*-(1→3)-β-D-Glc*p*-(1→4)-α-D-Glc*p*-(1↔1)-α-D-Glc*p*	3-,4-, and 6-hydroxyls of terminal Glc*p* unit of terminal trehalose	6,7,12
M. malmoense	LOS II	α-D-Man*p*-(1→3)-α-D-Man*p*-(1→2)-α-L-Rha*p*-(1→2)-[α-L-3-*O*-Me-Rha*p*-(1→2)]₂-α-L-Rha*p*-(1→3)-α-D-Glc*p*-(1↔1)-α-D-Glc*p*	3-,4-, and 6-hydroxyls of terminal Glc*p* unit of terminal trehalose	14
M. szulgai	LOS-I	α-L-2-*O*-Me-Fuc*p*-(1→3)-α-L-Rha*p*-(1→3)-β-L-Rha*p*-(1→3)-β-D-Glc*p*-(1→6)-α-D-Glc*p*-(1↔1)-α-D-2-*O*-Me-Glc*p*	3-,4-, and 6-hydroxyls of terminal 2-*O*-Me-Glc*p* unit of terminal mono-*O*-Me-trehalose	13

M. "linda"	LOS	β-D-Glcp-(1→3)-α-L-Rhap-(1→3)-α-D-Glcp-(1→3)-α-D-Glcp-(1↔1)-α-D-Glcp	3,4, and 6-hydroxyls of terminal Glcp unit of terminal trehalose	17
M. butyricum	LOS-I	4,6-(methyl 1′-carboxyethylidene)-3-O-Me-β-D-Glcp-(1→3)-4,6-(methyl 1′-carboxyethylidene)-β-D-Glcp-(1→4)-β-D-Glcp-(1→6)-α-D-Glcp-(1↔1)-α-D-Glcp	2′ and 3,4-hydroxyls of terminal trehalose	10
M. tuberculosis Canetti	LOS-I	*N*-acyl-4-amino-4,6-dideoxy-Galp-(1→4)-2-O-Me-α-L-Fucp-(1→3)-β-D-Glcp-(1→3)-2-O-Me-α-L-Rhap-(1→3)-2-O-Me-α-L-Rhap-(1→3)-β-D-Glcp-(1→3)-4-O-Me-α-L-Rhap-(1→3)-6-O-Me-α-D-Glcp-(1↔1)-α-D-Glcp	2,3,6 and 3,4,6-hydroxyls of terminal trehalose in the proportions of 2:3	15
M. gordonae 989	LOS-I	*N*-acyl-4-amino-4,6-dideoxy-2,3-O-Me-α-Galp-(1→3)-2-O-Me-4-O-Ac-α-L-Fucp-(1→3)-β-D-Glcp-(1→3)-2-O-Me-α-L-Rhap-(1→3)-[β-D-Xylp-(1→2)]-α-L-Rhap-(1→3)-β-D-Glcp-(1→3)-6-O-Me-α-D-Glcp-(1↔1)-α-D-Glcp	2,3,4 and 6-hydroxyls of terminal trehalose	16
M. gordonae 990	LOS-I	α-L-Rhap-(1→2)-3-O-Me-α-L-Rhap-(1→3)-[β-D-Xylp-(1→2)]-α-L-Rhap-(1→3)-β-D-Glcp-(1→3)-β-D-Glcp-(1→3)-α-L-Rhap-(1→3)-6-O-Me-α-D-Glcp-(1↔1)-α-D-Glcp	2,3,4 and 6-hydroxyls of terminal trehalose	16
M. fortuitum biovar *fortuitum*		β-D-Glcp-(1→6)-α-D-Glcp-(1↔1)-α-D-Glcp	2′ and 2,3,6-hydroxyls of terminal trehalose	18
M. tuberculosis H37Rv	DAT₁, DAT₂	α-D-Glcp-(1→1)-α-D-Glcp	2 and 3 hydroxyls of terminal trehalose	22

*a*ND = Not determined

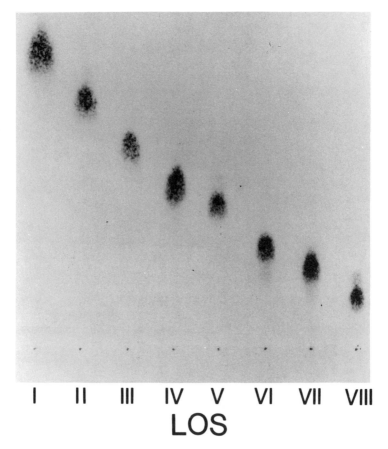

I   II   III   IV   V   VI   VII   VIII
LOS

Figure 2. Thin-layer chromatogram of the individual purified LOS I-VIII from *M. kansasii*. Solvent, CHCl$_3$-CH$_3$OH-H$_2$O (10:5:1). All plates were sprayed with 10% H$_2$SO$_4$ and heated.

dimethyltetradecanoic acid esterified the hydroxyl groups at C-3, C-4, and C-6 of the terminal glucosyl residue (Table I).

The species-specific glycolipids of *M. szulgai* which were also recognized as members of the LOS class required special hydrolytic conditions, combined with GC/MS for full structural elucidation, and this allowed the recognition of a 2-*O*-methyl-α-D-glucopyranosyl unit, which, in turn, led to the recognition of the presence of glycosidically linked 2-*O*-methyltrehalose (*13*). The complete structure of the oligosaccharide from the simplest member of the LOS family in *M. szulgai* has been elucidated (Table I).

M. malmoense, an atypical mycobacterium implicated in pulmonary infections, was also found to contain species-specific LOS glycolipids (*14*). De-*O*-acylation, followed by partial acidic cleavage to release the glycosidically-linked trehalose, α-mannosidase, *O*-deuteriomethylation, partial acid hydrolysis, reduction with NaB^{2}H$_4$, and *O*-ethylation, combined with ^{1}H-NMR, GC/MS and FAB/MS, revealed the structure of the major oligosaccharide as α-D-Man*p*-(1→3)-α-D-Man*p*-(1[2])-α-L-Rha*p*-(-]$_4$3)-α-L-Rha*p*-(1→3)]-α-D-Glc*p*-(1↔1)-α-D-Glc*p*, in which two of the 2-α-L-Rha*p* residues are *O*-methylated at C-3 (Table I). The distinct chromatographic mobility of the major LOS allowed one to distinguish *M. malmoense sensu stricto* from other closely, related, clinically significant mycobacteria.

Mycobacterium tuberculosis (strain Canetti) was found to be characterized by the presence of two novel LOS glycolipids (*15*). The structure of the major glycolipid was established (Table I) and was found to be unique in that the trehalose substituent was methylated at the 6'-position. In addition, unlike other well characterized members of the LOS family, which are invariably acylated at the 3,4, and 6 (or 4' and 6) positions, the product from *M. tuberculosis* Canetti consists of a mixture of 2,3,6- and 3,4,6-tri-*O*-acylated oligosaccharides.

Two clinical isolates of *Mycobacterium gordonae* were also found to contain a novel series of glycolipids based on the alkali-labile LOS. Inherent features of interest included a novel branching of the oligosaccharide backbone in the form of a single residue of terminal (t)-β-D-Xyl*p*, (t)-β-D-3-*O*-CH$_3$-Xyl*p*, or (t)-a-D-Ara*f*, and an incompletely defined *N*-acyl derivative of 4- amino-4,6-dideoxy-2,3-*O*-CH$_3$-galactopyranose residue in some members. Also of interest was a novel mono-6'-*O*-CH$_3$-2,3,4,6-tetra-*O*-acyl-glucose unit, representing the first example of such a unit in the LOS family of glycolipids from mycobacteria (*16*).

A smooth variant of a *Mycobacterium sp.* originating in a patient with Crohn's disease (*Mycobacterium "linda"*) provided a very simple version of the trehalose-containing LOS (*17*). The structure of the oligosaccharide unit in the one major LOS in this organism was identified as β-D-Glc*p*-(1→3)-α-L-Rha*p*-(1→3)-α-D-Glc*p*-(1↔1)-α-D-Glc*p*, and GC/MS allowed identification of the fatty acyl esters as primarily 2,4-dimethyltetradecanoate (Table I). The use of a mild *O*-methylation technique followed by a Hakomori *O*-ethylation allowed the ready location of three such acyl residues on the C-3, C-4, and C-6 positions of the terminal glucosyl residue of the trehalose unit (*17*).

Mycobacterium fortuitum, biovar *fortuitum*, the cause of serious skin and soft-tissue infections, was found to be differentiated from *M. fortuitum*, biovar *peregrinum*, and other rapidly growing opportunistic mycobacteria by the presence of yet another unique antigenic LOS glycolipid (Table I) (*18*).

Isolates of the tubercle bacillus are generally devoid of these specialized LOS glycolipids and their corresponding unusual sugars; the Canetti strain (Table I) seems to be an aberration in this respect. They do, however, have simple acyl trehaloses, quite apart from the mycolyltrehaloses (cord factor) and sulpholipids. Two distinct classes of such acylated trehaloses are the non-polar class, heavily acylated with mycolipenic and other long-chain acids, and the polar class, acylated with methyl-branched, hydroxy mycolipanolic, the methyl-branched mycocerosic and straight-chain fatty acids (*19, 20, 21*). This latter class has now been shown to be a complex mixture of 2,3-di-*O*-acyltrehaloses (*22*) (Table I).

The structures of the trehalose-containing lipooligosaccharides described to date are summarized in Table I.

The Phenolic Glycolipids of Mycobacteria

Early work documented in Brennan (*3*) resulted in the isolation of a glycolipid from *M. bovis* having a characteristic IR spectrum. This glycolipid was termed mycoside B, the name mycoside being defined as a "type-specific glycolipid of mycobacterial origin." This type of mycoside is more correctly termed a glycosylphenolphthiocerol dimycocerosate and was later simply termed phenolic glycolipids by Hunter and Brennan (*23*). The generic structure of this class of glycolipids is shown in Figure 1, and the attached glycosyl units recognized to date are described in Table II. Phenolic glycolipid I of *M. leprae* and related neoantigens are widely used in the serodiagnosis of leprosy (*5*). This glycolipid is also implicated in the pathogenesis of leprosy, the emergence of suppressor thymocyte cells (*24*) and, through its ability to scavenge oxygen radicals, in the intracellular survival of *M. leprae* within macrophages of individuals with lepromatous leprosy (*25*). The triglycosyl phenolphthiocerol dimycocerosate of *M. tuberculosis*, PGL-Tb1, is apparently confined to certain smooth, avirulent strains and is apparently not present in virulent strains of *M. tuberculosis* and therefore, unlike PGL-I from *M. leprae*, is not useful in the serodiagnosis and management of tuberculosis (*26*).

The structure of the major phenolic glycolipids described to date are summarized in Table II.

The Glycopeptidolipids of Mycobacteria

The historical background of another class of mycobacterial glycolipids, the C-mycoside glycopeptidolipids (Figure 1), has been carefully reviewed by Goren (*27*). By seroagglutination of whole cells, Schaefer subdivided the *M. avium*, *M. intracellulare*, *M. scrofulaceum* group (MAIS) into 31 distinct serotypes. The detailed oligosaccharide structures of the GPLs from 12 of these serovars have been completely defined (Table III) (see ref. *28*). The terminal sugars of the

Table II Structures of Major Phenolic Glycolipids from *Mycobacterium*

Species	Trivial Name	Structure of Oligosaccharide	Ref.
M. bovis	Mycoside B	2-*O*-Me-α-L-Rhap-(1→phenol dimycocerosyl phthiocerol)	2, 68, 69
M. bovis	Mycoside B'	α-L-Rhap-(1→phenol dimycocerosyl phthiocerol)	68
M. bovis	Mycoside B''	2-*O*-Me-α-L-Rhap-(1→phenol phthiocerol)	68
M. leprae	PGL-I	3,6-di-*O*-Me-β-D-Glcp-(1→4)-2,3-di-*O*-Me-α-L-Rhap-(1→4)-3-*O*-Me-α-L-Rhap-(1→phenol dimycocerosyl phthiocerol)	70, 71
M. kansasii	PGL-KI (Mycoside A)	2,6-dideoxy-4-*O*-Me-α-*arabino*-Hexp-(1→3)-4-*O*-Ac-2-*O*-Me-α-L-Fucp-(1→3)-2-*O*-Me-α-L-Rhap-(1→3)-2,4-di-*O*-Me-α-L-Rhap-(1→phenol dimycocerosyl phthiocerol)	72, 73
M. tuberculosis Canetti	PGL-Tb1	2,3,4-tri-*O*-Me-α-L-Fucp-(1→3)-α-L-Rhap-(1→3)-2-*O*-Me-α-L-Rhap-(1→phenol dimycocerosyl phthiocerol)	69
M. haemophilum	--	2,3-di-*O*-Me-α-L-Rhap-(1→2)-3-*O*-Me-α-L-Rhap-(1→4)-2,3-di-*O*-Me-α-L-Rhap-(1→phenol dimycocerosyl phthiocerol)	61
M. ulcerans	--	3-*O*-Me-α-L-Rhap-(1→phenol dimycocerosyl phthiocerol)	74
M. marinum	Mycoside A	3-*O*-Me-α-L-Rhap-(1→phenol dimycocerosyl phthiocerol)	75, 76
M. gastri	PGL-GI	2,6-dideoxy-4-*O*-Me-α-*arabino*-Hexp-(1→3)-4-*O*-Ac-2-*O*-Me-α-L-Fucp-(1→3)-2-*O*-Me-α-L-Rhap-(1→3)-2,4-di-*O*-Me-α-L-Rhap-(1→phenol dimycocerosyl phthiocerol)	77

Table III. Structures of the Major Oligosaccharide Haptens from Glycopeptidolipid Antigens from *Mycobacterium*

Species	Serovar No.	Structure of Oligosaccharide	Reference
M. avium	1	O-Ac-α-L-Rhap-(1→2)-6deoxy-L-Tal	3, 27, 29
	2	4-O-Ac-2,3-di-O-Me-α-L-Fucp-(1→3)-α-L-Rhap-(1→2)-6deoxy-L-Tal	
	9	4-O-Ac-2,3-di-O-Me-α-L-Fucp-(1→4)-β-D-GlcAp-(1→4)-2,3-di-O-Me-α-L-Fucp-(1→3)-α-L-Rhap-(1→2)-6deoxy-L-Tal	
	4	4-O-Me-α-L-Rhap-(1→4)-2-O-Me-α-L-Fucp-(1→3)-L-Rhap-(1→2)-6deoxy-L-Tal	
	20	2-O-Me-α-D-Rhap-(1→3)-2-O-Me-α-D-Rhap-(1→3)-2-O-Me-α-L-Fucp-(1→3)-α-L-Rhap-(1→2)-6deoxy-L-Tal	
	14	4-formamido-4,6-dideoxy-2-O-Me-3-C-Me-α-Manp-(1→3)-2-O-Me-α-D-Rhap-(1→3)-2-O-Me-α-L-Fucp-(1→3)-α-L-Rhap-(1→2)-6deoxy-L-Tal	
	25	4-acetamido-4,6-dideoxy-2-O-Me-α-FucNAc-(1→4)-β-D-GlcAp-(1→4)-2-O-Me-α-L-Fucp-(1→3)-α-L-Rhap-(1→2)-6deoxy-L-Tal	
	26	2,4-di-O-Me-α-L-Fucp-(1→4)-β-D-GlcAp-(1→4)-2-O-Me-α-L-Fucp-(1→3)-α-L-Rhap-(1→2)-6deoxy-L-Tal	
	8	4,6-(1'-carboxyethylidene)-3-O-Me-β-D-Glcp-(1→3)-α-L-Rhap-(1→3)-α-L-Rhap-(1→2)-6deoxy-L-Tal	
	21	4,6-(1'-carboxyethylidene)-β-D-Glcp-(1→3)-α-L-Rhap-(1→3)-α-L-Rhap-(1→2)-6deoxy-L-Tal	
	12	4-(2'-hydroxy)propionamido-4,6-dideoxy-3-O-Me-β-D-Glcp-(1→3)-4-O-Me-α-L-Rhap-(1→3)-α-L-Rhap-(1→2)-6deoxy-L-Tal	
	17	3-(2'-methyl)(3'-hydroxy)butyramido-3,6-dideoxy-β-D-Glcp-(1→3)-4-O-Me-α-L-Rhap-(1→3)-α-L-Rhap-(1→2)-6deoxy-L-Tal	
	19	3,4-di-O-Me-β-D-GlcAp-(1→3)-2,4-di-O-Me-3-C-Me-6deoxy-α-L-Rhap-(1→3)-α-L-Rhap-(1→2)-6deoxy-L-Tal	
M. fortuitum biovar *peregrinum*		3,4-di-O-Me-α-L-Rhap-(1→2)-3,4-di-O-Me-α-L-Rhap or 3-O-Me-α-L-Rhap-(1→2)-3,4-di-O-Me-α-L-Rhap	39
M. senegalense M263		3,4-di-O-Me-α-L-Rhap-(1→2)-3,4-di-O-Me-α-L-Rhap or 3-O-Me-α-L-Rhap-(1→2)-3,4-di-O-Me-α-L-Rhap	40
M. xenopi CIPT 140 35004		2,3,4-tri-O-Me-α-L-Rhap-(1→3)-α-L-Rhap-(1→3)-α-L-Rhap-(1→3)-α-L-6deoxy-Glcp	42, 43
M. xenopi ATCC 19250		α-L-Rhap-(1→3)-α-L-Rhap	44
M. paratuberculosis		4-O-Ac-2,3-di-O-Me-α-L-Fucp-(1→3)-α-L-Rhap-(1→2)-6deoxy-L-Tal	35

oligosaccharide haptens were found to contain a rich array of novel amido sugars, branched-chain sugars, sugar acids, and pyruvylated sugars (*29*). The use of monoclonal antibodies to the individual serovars of the *M. avium* complex in conjunction with semisynthetic neoantigens containing some of the precise terminal sugar combinations (*30-33*) have established the antigenic dominant epitopes of individual serovars (*34*). The glycopeptidolipids are not exclusively confined to the MAIS organisms; they are also observed in other mycobacterial species such as *M. paratuberculosis* (*35, 36*), *M. chelonae* subsp. *chelonae* (*37*), and *M. simiae* I, II (*38*).

More recently, the structures of the major C-mycoside GPLs from *M. fortuitum* biovar. *peregrinum* have been established (*39*). These compounds share a peptidolipid core linked to saccharide units as found in the C-mycoside GPLs of *M. avium*, but differ in the distribution and nature of the sugar residues leading to previously undefined C-mycoside structures. These structures are unusual in that they do not contain 6-deoxytalose or its derivatives; moreover, the oligosaccharide portion is linked to the alaninol residue instead of the *allo*-threonine (*39*). A pair of glycolipids from a strain of *M. senegalense* (*40*) were found to, in fact, correspond to those described by Lopez-Marin *et al.* (*39*), from *M. fortuitum* biovar. *peregrinum*. Based on the observations of Jenkins *et al.* (*41*) that *Mycobacterium xenopi* should be assimilated into the *M. avium* complex, Riviere and Puzo (*42, 43*) reported that a strain of *M. xenopi* contained an unexpected, unique variant of the GPL structure (a non-C-mycoside GPL) in which the core consisted of fatty acyl-NH-L-Ser-L-Ser-L-Phe-*allo*-D-Thr-OCH$_3$ rather than the expected lipopeptide core. The oligosaccharide hapten is linked glycosidically to the *allo*-Thr-OCH$_3$ residue and consisted of 2,3,4-tri-*O*-CH$_3$-α-L-Rha*p*-(1→3)-2-*O*-lauryl-α-L- Rha*p*-(1→3)-α-L-Rha*p*-(1→3)-2,4-di-*O*-(acetyl, lauryl)-L- 6dGlc*p*. The remaining 3-*O*-CH$_3$-6dTal*p* was glycosidically linked to the distal serine residue. Recent studies by Besra *et al.* (*44*) support the essence of the GPL structure described by Riviere and Puzo (*42, 43*) but suggested important modifications in both the lipopeptide core and sugar appendages, which, in turn, suggest that *M. xenopi*, like *M. avium*, exists as a serocomplex in nature. The structures of the glycopeptidolipids described to date are summarized in Table III.

Purification and Protocols in the Analysis of Species-Specific Glycolipids.

The following is an account of the experimental approach leading to the elucidation of most of those structures, at least those arising in this laboratory (*28*).

In general, cultures are grown for 6-8 weeks at 37°C in Fernbach flasks containing 7H11 medium with gentle agitation on a shaker. Both cells and medium are harvested by autoclaving and their entire contents evaporated to dryness. The resulting solids are extracted twice with CHCl$_3$-CH$_3$OH (2:1) at 50°C overnight, dried, and partitioned between the aqueous and organic phases arising from a mixture of CHCl$_3$-CH$_3$OH-H$_2$O (4:2:1) (*45*). The contents of the lower organic phase are dried and applied to a column (2.5 x 30 cm) of Florisil

(100-200 mesh) (Fisher Scientific Co., Pittsburgh, PA) and irrigated with 500 ml each of $CHCl_3$, as described (7). Eluates (10 ml) are collected and subjected to TLC in $CHCl_3$-CH_3OH-H_2O (90:10:1; 30:8:1; or 65:25:4) and sprayed with 10% sulfuric acid in ethanol followed by heating at 110°C for 5 min. Glycolipid containing fractions are pooled and further purified by preparative TLC on silica gel (Merck 5735 silica gel 60F_{254}, Darmstadt, Germany) in the above solvent systems. To aid in the isolation of specific GPLs, the crude lipid extract is dissolved in $CHCl_3$-CH_3OH (2:1, 30 ml/g) and treated with an equal volume of 0.2 M NaOH for 35 min at 40°C, neutralized with glacial acetic acid, dried, and the residue suspended in $CHCl_3$-CH_3OH-H_2O (4:2:1). The organic layer is separated, dried, and chromatographed as described above. However, this procedure also results in deacetylation/deacylation of acetylated/acylated GPLs which may have untoward effects on the antigenicity if the O-acetyl group/O-acyl group is located on the terminal sugar epitope (34). Procedures for the isolation of naturally O-acetylated/O- acylated GPLs by HPLC have been described (46). Pure LOSes are used as the source of corresponding oligosaccharide (Ose); some clearly resolvable LOSs yield the same Ose (see Table I). Typically, 30 mg of pure LOS are dissolved in $CHCl_3$-CH_3OH (2:1; 1 ml) and reacted with an equal volume of 0.2 N NaOH in CH_3OH at 37°C for 1 h. The reaction mixture is neutralized with glacial acetic acid, dried, the products partitioned between $CHCl_3$ and H_2O, and the $CHCl_3$ phase backwashed twice with H_2O. The combined aqueous phases, which contain the neutral oligosaccharides, are purified by gel filtration on a column (1 x 175 cm) of BioGel P-2 in water. The lower organic phase from the above separation serves as a source of fatty acids which were examined by GC/MS (22) as methyl esters.

Composition and Absolute Configuration of Glycosyl Residues. Purified glycolipids were hydrolyzed in 2 M trifluoroacetic acid (TFA) at 120°C for 2 h as described (46). Glycosyl residues are reduced with NaB^2H_4 and the resultant alditols O-acetylated and examined by GC and GC/MS (46). To establish absolute sugar configurations, purified glycolipids were hydrolyzed in 1 M HCl in (R)- (-)-2-butanol (Aldrich Chemical Co., Milwaukee, WI), trimethylsilylated with TRI-SIL (Pierce Chemical Co., Rockford, IL), and the trimethylsilyl (R)-(-)-2-butylglycosides examined by GC/MS (47) and compared to authentic standards.

Amino Acid Analysis. Amino acids were analyzed as their N(O)-heptafluorobutyryl-butyl esters [N(O)-HFB] (48). GPLs were completely hydrolyzed with 200 µl of 6 N HCl (Pierce Chemical Co., Rockford, IL) at 110°C for 12 h. Upon evaporation to dryness, the resulting reaction mixture was treated with 100 µl of 3 M HCl in butanol [isobutanol, or, to determine the absolute configuration (R)-(-)-2-butanol] at 120°C for 20 min. Upon careful drying, 100 µl of anhydrous ethyl acetate and 40 µl of heptafluorobutyric anhydride (Aldrich Chemical Co., Milwaukee, WI) was added, and the reaction mixture was heated at 150°C for 5 min and subsequently analyzed by GC/MS as described (49).

Alkylation of Glycolipids and Glycosyl Linkage Analysis. Glycolipids (1-2 mg) were suspended is 0.5 ml of dimethyl sulfoxide (Pierce Chemical Co., Rockford, IL), 100 μl of 4.8 M dimethyl sulfinyl carbanian added (*50, 51*), and the reaction mixture was stirred for 1 h. The alkylating reagent (50 μl) was added slowly and the suspension stirred overnight. The reaction mixture was then diluted with 0.5 ml water and the resulting product applied to a C-18 Sep-Pak cartridge (Waters, Milford, MA), as described (*52*). The O-methylated glycolipid which appeared in the acetonitrile eluant was hydrolyzed with 2 M TFA at 120°C for 1 h (*46*). The resulting hydrolysate was reduced with NaB^2H_4, O-acetylated, and examined by GC/MS.

In order to establish the location of acyl functions on an oligosaccharide backbone, the native glycolipid was subjected to the neutral alkylating conditions of Prehm (*53*), as follows. To the pure glycolipid (2 mg), under N_2, was added 30 μl of 2,6-di-*tert*-butylpyridine, 20 μl methyltrifluoromethanesulfonate, and 200 μl of trimethylphosphate. The reaction mixture was stirred at room temperature for 5 h, following which 1 ml of water was added. The mixture was applied to a C-18 Sep-Pak cartridge (Waters) as described (*52*). The ethanol eluant was dried to yield the naturally acylated, O-methylated glycolipid, which was further methylated with C^2H_3I using the Hakomori methylation procedure (*51*) and purified on a C-18 Sep-Pak cartridge (Waters) as described (*54*). The O- trideuteriomethylated, O-methylated oligosaccharide was recovered in the acetonitrile eluant and subjected to acid hydrolysis in 2 M TFA at 120°C for 1 h (*46*). The resulting hydrolysate was reduced with NaB^2H_4, O-acetylated, and examined by GC/MS. In some cases, the O-trideuteriomethylated, O-methylated oligosaccharides were partially hydrolyzed with 2 M TFA at 90°C for 1 h to generate smaller, partially O-alkylated oligosaccharide fragments. The hydrolysates were dried, reduced with NaB^2H_4, and re-methylated with CH_3I (*55*), and the resultant products recovered by chromatography on a C-18 Sep-Pak cartridge as described (*54*). The resultant oligosaccharide fragments were partially fractionated by HPLC as described previously and analyzed by GC/MS (*46*).

Preparation of Oligosaccharide Haptens of Individual GPLS as the Oligosaccharide Alditols. The original procedure for reductive β-elimination (*56*) has been improved to allow for greater recovery of the oligoglycosylalditol. Purified GPLs (5 mg) were heated to 65°C for 18 h with a mixture of 0.5 M NaOH and 1 M $NaBH_4$ in a total volume of 3 ml of ethanol/water (1:1; v/v). Excess borohydride was destroyed with acetic acid, evaporated, and boric acid eliminated as methyl borate by codistillation with methanol. The residue was dissolved in $CHCl_3$-CH_3OH-H_2O (4:2:1). The upper aqueous phase was separated and passed through a small column of Dowex 50(H^+) and then evaporated to dryness to yield the oligoglycosylalditol.

NMR Analysis. Routine ^1H, ^{13}C, 2D $^1H/^{13}C$, and 2D COSY NMR were obtained for native glycolipids in C^2HCl_3-$C^2H_3O^2H$ (2:1; 0.5 ml) before and after exchanging protons with C^2HCl_3-$C^2H_3O^2H$ (2:1). Spectra for pure oligosaccharides were obtained in ^2H_2O (0.5 ml) before and after exchanging protons with ^2H_2O.

Gas Chromatography-Mass Spectrometry, Fast-Atom Bombardment-Mass Spectrometry, and Plasma Desorption-Mass Spectrometry. GC/MS of O-alkylated, partially O-alkylated alditol acetates or N-(O)-HFB butyl esters were performed on a Hewlett-Packard 5890 gas chromatograph connected to a Hewlett-Packard 5790 mass selective detector as described previously (44). Gas chromatography was routinely conducted on a fused silica capillary column of Durabond-1 (J&W Scientific, Rancho Cordova, CA) as described (46).

FAB/MS was performed on a VG 7070 EHF mass spectrometer equipped with an Ion Tech saddle field gun operating at 7-8 kv and 1mA with xenon gas (44). Alternatively, a VG Analytical ZAB-HF mass spectrometer fitted with a M-Scan FAB gun operating at 10 kv was used (57). Samples were applied in either 50:50 glycerol-thioglycerol or M-nitrobenzyl alcohol matrix. Plasma desorption mass spectrometry (PDMS) of intact mycobacterial antigenic glycolipids was achieved as previously described (58). Glycolipids were O-acetylated using a 2:1 (v/v) mixture of trifluoroacetic anhydride/acetic acid as modified for microscale derivatization of glycolipids by Dell and colleagues (59, 60). Alternatively, O-deuterioacetylation was conducted in pyridine-hexadeuterioacetic anhydride (1:1) at 80°C for 2 h (59).

Structural Analysis of Mycobacterial Lipids

The following section discusses the structural analysis of examples of the GPL, LOS, and PGL families of glycolipids; however, each glycolipid often requires unique approaches in its structural elucidation. The complete structural analysis of a glycolipid may be broken down into several distinct stages: glycosyl composition, absolute configuration of glycosyl residues, glycosyl linkage analysis, anomeric configuration of each glycosyl residue, sequence of glycosyl residues, identification and location of fatty acyl groups, and identification and location of other noncarbohydrate substituents (28). Each of these stages is discussed in the context of mycobacterial glycolipids, of all three classes, whose structures have been elucidated to date, with an emphasis on the application of mass spectrometry to their elucidation (28).

Determination of Glycosyl Composition. A convenient method for the identification of glycosyl residues is GC, GC/MS of alditol acetates, produced after hydrolysis, reduction and O-acetylation (46, 52). This method is extremely useful due to the presence of endogenously methylated glycosyl residues which are found in mycobacteria and are readily identified by GC/MS by comparison with authentic standards. For instance, the PGL antigen from *M. haemophilum* which contains two residues of 2,3-di-O-CH_3-rhamnitol and a single residue of

3-*O*-CH$_3$-rhamnitol (Figure 3) are easily distinguished by their characteristic GC retention times and fragment ions. A disadvantage of this procedure is that uronosyl residues, in the case of GPL antigens from *M. avium* serovars 9 and 25, do not form suitable volatile derivatives for detection. However, this problem can be overcome by the application of various methods of carboxyl reduction (*61, 62*) prior to alditol acetate formation. Many of the novel glycosyl residues present in mycobacterial glycolipids, for example, the novel amino sugar in *M. tuberculosis* Canetti PGL, are degraded by 2 M TFA hydrolysis, and the use of anhydrous HF, mild HCl, or β-elimination is recommended.

Elucidation of Absolute Configuration of Glycosyl Residues, Ring Size, Linkage Arrangement Between Glycosyl Residues and Anomeric Configurations. The procedure usually employed involves comparative GC and GC/MS of volatile derivatives of optically pure 2-butylglycosides (*47*). In the case of new mycobacterial lipids which contain endogenous *O*-methylated sugars, the glycolipid must be de-*O*-methylated (*46*) or the methylated standard must be available. In some cases, chemical synthesis (*63*) must be undertaken as in the case of the 4-amino sugars present in *M. kansasii*.

Glycosyl linkage patterns and the question of ring size within mycobacterial glycolipids is routinely examined by conventional methylation analysis involving GC/MS (*50, 51*). The methodology is dependent on known fragmentation patterns of partially *O*-methylated alditol acetates (*64*). The derived alditol acetates from the *O*- methylated LOS from *M. fortuitum* biovar *fortuitum* (Figure 4) allowed the ready recognition of a single residue of 6-linked Glc*p* and two residues of terminal Glc*p*.

Anomeric configurations of glycolipids are routinely established by ^{1}H-NMR (*65*). However, in some cases, such as the *manno* configuration, which give rise to signals in the ^{1}H-NMR spectra between δ 4.7 and δ 4.9 which could be attributable to either a- or β-glycosyl linkages, anomeric configurations are determined by C-1/H-1 coupling constants (*66*).

Sequence of Glycosyl Residues. FAB/MS and Californium-252 plasma desorption (PD)/MS have proved to be powerful tools for sequence determination (*58*). The complete structure of the oligosaccharide moiety of LOS-II (Ose-II) of *M. tuberculosis* Canetti was analyzed by positive-ion FAB/MS (Figure 5). The (M+Na)$^+$ ion observed at *m/z* 1673 and *m/z* 1657 (M+Li)$^+$ with the addition of LiCl (results not shown), clearly demonstrated a molecular weight of 1650 for Ose-II. The sequence of the oligosaccharide was deduced as shown in Figure 9. The J$_2$ trehalose-containing ions were found in two forms, J$_2$ + OH + Na$^+$ and J$_2$ + OH + H$^+$ (*67*). The smallest of this series of ions, those at *m/z* 517 (J$_2$ + OH + H$^+$) and *m/z* 539 (J$_2$ + OH + Na$^+$) correspond to the trisaccharide containing a mono-*O*-CH$_3$-trehalose. However, problems do arise when glycolipids contain different glycosyl residues of the same weight or, alternatively, when the same glycosyl residue is present in different linkage forms whereby alternative structures yield the same mass values. This was, in fact, true for Ose-II from *M. tuberculosis* Canetti where three of the glycosyl residues were

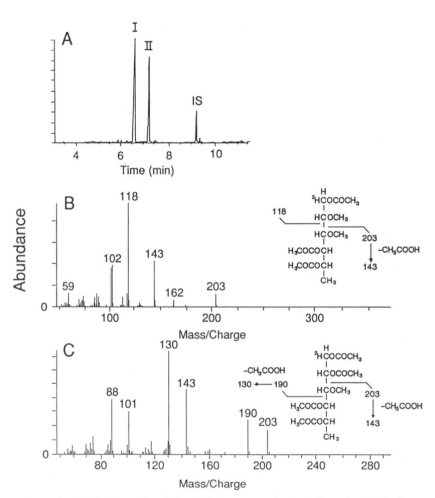

Figure 3. GC/MS profile of the alditol acetates derived from the PGL of
M. haemophilum. (A). Total ion chromatogram; I, 2,3-di-*O*-CH₃-rhamnitol
(R_T 6.58 min); II, 3-*O*-CH₃-rhamnitol (R_T 7.20 min); IS, inositol. (B). Mass
spectrum of (I) 2,3-di-*O*-CH₃-rhamnitol showing characteristic fragment ions.
(C). Mass spectrum of (II) 3-*O*-CH₃-rhamnitol showing characteristic
fragment ions.

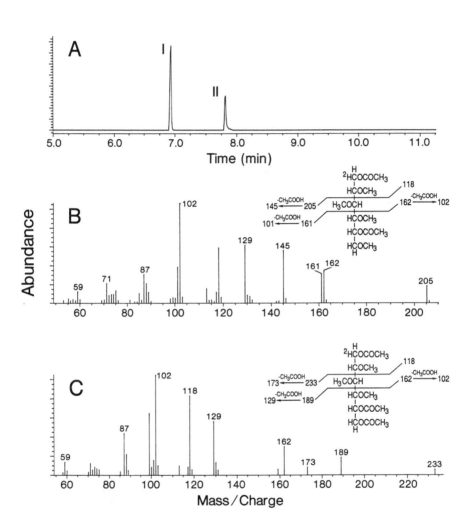

Figure 4. GC/MS profile of the alditol acetates derived from the O-methylated oligosaccharide of the LOS of *M. fortuitum*. (A). Total ion chromatogram; I, 1,5-di-O-CH₃O-2,3,4,6-tetra-O-CH₃-glucitol (R_T 6.93 min); II, 1,5,6-tri-O-CH₃O-2,3,4-tri-O-CH₃-glucitol (R_T 7.82 min). (B). Mass spectrum of (I) showing characteristic fragment ions. (C). Mass spectrum of (II) showing characteristic fragment ions.

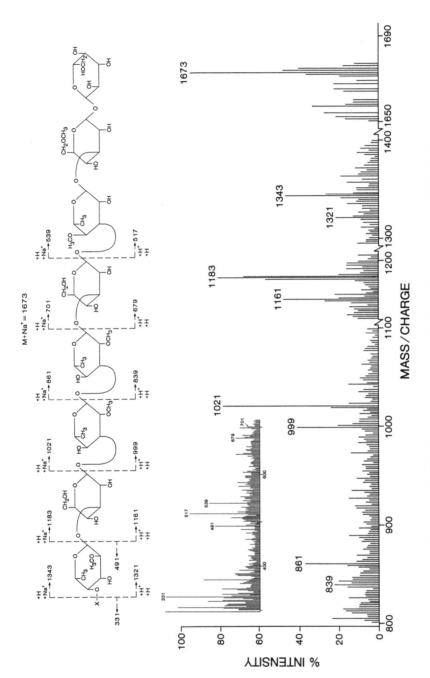

Figure 5. FAB-MS of the nonasaccharide (Ose-II) derived from LOS-II of *M. tuberculosis* Canetti. *x* = amino sugar residue.

4-O-CH$_3$-Rha and 2-O-CH$_3$-Rha (x 2). In order to confirm/complement the glycosyl sequence obtained from FAB/MS, a partial depolymerization of the oligosaccharide was carried out to generate smaller fragment oligosaccharides which were resolved by HPLC or GC, followed by mass spectrometry (*15*). The resulting fragment sequences thereby allow the construction of sugar sequences within the original polymer. The structural elucidation of a number of glycolipids from mycobacteria has successfully been resolved using this approach.

FAB/MS of deacylated LOSs produces important molecular weight and glycosyl sequences data (*7*); in contrast, FAB/MS of native LOSs produces very little information. However, positive ion PDMS analysis of the intact LOS of *M. malmoense* produced a molecular weight associated ion (M+Na)$^+$ at 2276. The negative ion PDMS spectrum of the same sample revealed the presence of carbohydrate anions of C$_8$, C$_{21}$ and C$_{27}$ fatty acyls. These fatty acyl functions were in accord with the deduced structure and molecular wieght of 2253.

Peracetylation of the *M. malmoense* LOS, followed by PDMS analysis, produced the spectrum shown in Figure 6, which resulted in the acquisition of further sequence data. The molecular weight-associated ion (M+Na)$^+$ was shifted to *m/z* 3074, an increase of 798 or 19 O-CH$_3$CO groups. As a consequence, extensive carbohydrate sequence was obtained, with sequence ions at *m/z* 331, 620, 850, 1052, 1254, 1484, 1715, and 2003, as well as 1033, which confirmed the sugar sequence and located all three fatty acyl groups to the terminal glucosyl residue of the trehalose unit. This methodology first suggested the presence of a C$_8$ fatty acyl group which had not been detected previously by GC/MS analysis of fatty acid methyl esters.

FAB/MS of the native PGL-I of *M. leprae* has met with very little success; however, negative- and positive-ion PDMS have provided very useful information (*58*). The negative-ion spectrum clearly demonstrated the presence of the characteristic mycocerosic acids (C$_{30}$, *m/z* 452; C$_{32}$, *m/z* 480; C$_{34}$, *m/z* 508). The positive-ion PDMS spectrum revealed a cluster of ions at *m/z* 2000-2100 and were designated as the C-30,34, the C-32,34, and the C34,34 species of PGL-I. Per-O-acetylation of PGL-I, followed by PDMS analysis, resulted in an increase of 126 or three O- CH$_3$CO groups. A similar shift was noted for the sugar fragments of the PGL-I (*m/z* 191→275; *m/z* 365→447; *m/z* 526→652). Clearly, the *M. leprae* PGL-I is a complex mixture, and the use of PDMS has proved to be a valuable analytical tool in the structural definition of this important glycolipid.

Identification and Location of Fatty Acyl Groups and Other Noncarbohydrate, Nonfatty Acyl Groups. Fatty acyl groups are readily established by GC and GC/MS of methyl esters (*22*). Other lipid groups such as phthiocerols, as illustrated for *M. haemophilum* (Figure 7), are released from the glycolipid by solvolysis and analyzed by GC/MS (*61*).

Fatty acyl groups are preferentially located using the mild alkylating conditions of Prehm (*53*) under which the fatty acyl groups are stable and protect the glycosyl oxygens from the addition of methyl groups. This is followed by a Hakomori methylation (*51*) which replaces the fatty acyl groups with a C^{2}H$_3$ or

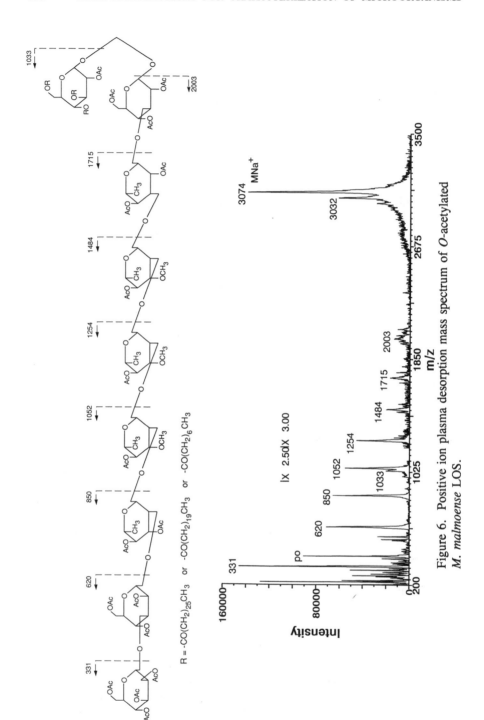

Figure 6. Positive ion plasma desorption mass spectrum of O-acetylated M. malmoense LOS.

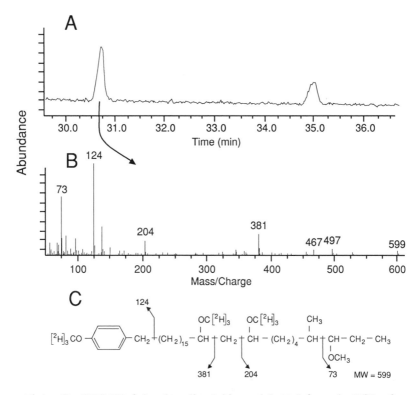

Figure 7. GC/MS of the phenolic phthiocerol "core" from the PGL of *M. hemophilum*. The deacylated phenolphthiocerol A moiety was *O*-trideuteriomethylated and examined by GC/MS. (A). Total ion chromatogram. (B). Mass spectrum of major components. (C). An illustration of the formation of the major fragment ion.

C_2H_5 group depending on the alkylating reagent used. The O-alkylated oligosaccharide is then hydrolyzed, reduced, with NaB^2H_4 and O-acetylated (46). A close examination of the resultant alditol acetates enables one to establish the position of the labelled C^2H_3 or C_2H_5 introduced during the Hakomori methylation and thus the position of acylation.

Determination of noncarbohydrate, non-fatty acyl groups is most relevant to the GPLs, since they contain a characteristic array of amino acids and the amino alcohol L-alaninol. Standard amino acid and peptide analysis and mass spectrometry are now used to confirm the existence of these moieties (29). The GPLs of M. xenopi (42, 43, 44) yield an unusual amino acid pattern, different from those of the conventional GPLs of M. avium. GC and GC/MS analyses of the $N(O)$-HFB butyl esters of the amino acids of GPL-IV (Figure 8) revealed four amino acids, three of which were readily identified as Ser, alloThr and Phe (44). The fourth derivative was identified as 3-O-CH_3-Ser by comparative GC/MS (Figure 9). The enantiomeric configuration of the amino acids was established by comparing the GC profile of the $N(O)$-HFB (R)-2-butyl esters of the amino acids in GPL-IV with the corresponding derivatives of authentic standards. The 3-O-CH_3.Ser, Ser and Phe are in the L-configuration, whereas the alloThr is in the D-configuration (44).

Fast-Atom Bombardment Mass Spectrometry Analysis. FAB/MS has the capability of yielding molecular ions and adduct ions $(M + Na)^+$, as well as sequence ions, and has been applied extensively to the newer mycobacterial lipids. These mass spectral methods in conjunction with NMR provide the basic information on glycosyl composition, glycosyl sequences, identification and location of fatty acyl and other noncarbohydrate groups. FAB/MS analysis of GPLs, using either a 1:1 glycerol:thioglycerol or M- nitrobenzyl alcohol matrix doped with an alkali salt, usually affords an intense molecular ion from which the molecular weight of the intact GPL can be determined. Also, sequence information on the peptide and oligosaccharide moieties can be obtained by linked scanning, as recently discussed by Promé and co-workers (78). For instance, FAB/MS analysis of the O- trideuterioacetylated GPL-II (Figure 9A) and GPL-IV (Figure 9B) of M. xenopi (44) yielded valuable molecular weight information and readily interpretable fragmentation patterns. Assignments of the major signals are summarized in Figure 10. The A-type ion at m/z 248 in the spectra of GPL-IV arose from a mono-O-CH_3, mono-O-CH_3CO, mono-O-C^2H_3CO-6-deoxyhexosyl residue, and, since the only O-CH_3-sugar present in GPL-IV is 3-O-CH_3-6-dTal, which is acylated at position 2, the ion at m/z 248 points to the presence of an O-CH_3CO function. The ions at m/z 391 and m/z 363 are consistent with a di-O-C^2H_3CO-mono-O-acyl Rha residue in which the acyl function is C-10 (m/z 391) and C-8 (m/z 363). The methylation analysis demonstrated that this acyl function was at C-4 of this Rha unit. A-type fragment ions (Figure 9B) were seen at m/z 764 and m/z 736, indicative of the presence of dirhamnosyl units as

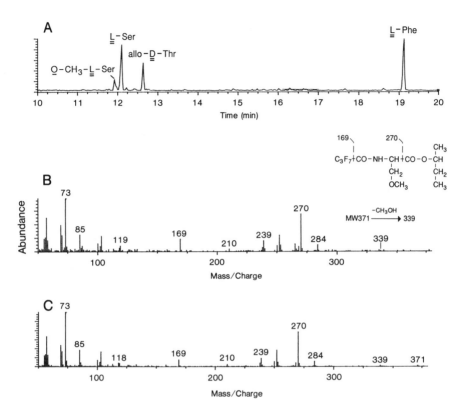

Figure 8.   The GC/MS profile of the amino acid *N(O)*-heptafluorobutyryl (R)-2-butyl esters derived from PGL-IV. (A). Total ion chromatogram. (B). Mass spectrum of the *O*-CH₃-Ser from GPL-IV. (C). Mass spectrum of the standard 3-*O*-CH₃-Ser.

opposed to a combination of 6-dTal and Rha. These ions also demonstrate that the reducing Rha residue is substituted with a C-12 fatty acid. The methylation data already demonstrated that this acyl residue must be on C-2.

Fragment ions which allowed the definition of the peptide sequence and the attachment points of the mono- and disaccharides were also found as shown in Figure 10. Thus, the use of the analytical protocols described in this review, in conjunction with various mass spectral techniques, allowed the complete structural definition of this new member of the GPL class of glycolipids.

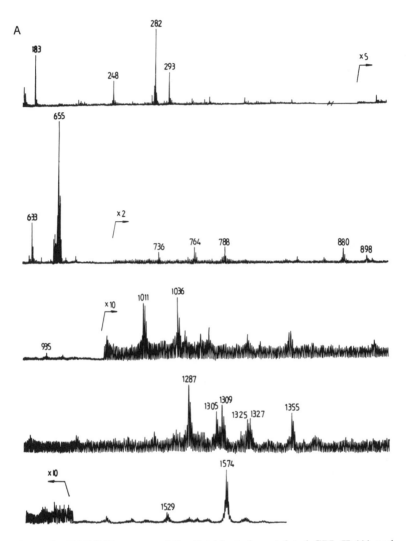

Figure 9. FAB/MS spectra of the *O*-trideuterioacetylated GPL-II (A) and GPL-IV (B). Major signals were assigned as shown in Figure 10. Signals at *m/z* 1309 and 1327 in (A) are the sodiated analogues of *m/z* 1287 and 1305 respectively, while the signal at *m/z* 1355 (28 amu higher than *m/z* 1327) corresponds to a similarly sodiated fragment ion resulting from ring cleavage of the 6- deoxytalose residue. The signals at *m/z* 1011 (A) and 1092/1120 (B) are 25 mass units below *m/z* 1036 (A) and 1117/1145, respectively. It is possible that these signals are sodiated C-terminal fragment ions resulting from cleavage between the amide nitrogen and the a-carbon of the methylated serine concomitant with loss of methoxide from the side-chain. The small signals at *m/z* 736 and 764 in (A) were probably derived from minor higher molecular weight components. The signals at *m/z* 633 and *m/z* 651 were derived from loss of both saccharide moieties from the peptide core.

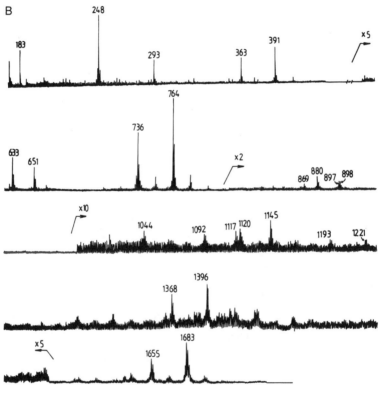

Figure 9. Continued.

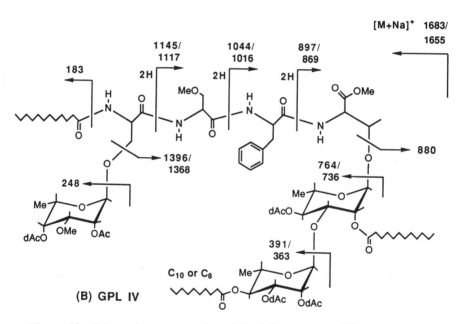

Figure 10. Schematic representation of the full structures of the
O-deuterioacetylated GPL-II (A) and GPL-IV (B). The major ions observed
in their FAB-MS spectra (Figure 9) are indicated. dAc, deuterioacetyl; Ac,
acetyl; Me, methyl.

Acknowledgments

We thank Becky Rivoire, Kim Robuck, and Russ Suzuki for valuable technical help and Dr. Michael McNeil for help with experimental design. We thank Marilyn Hein for preparation of the manuscript. The research reported here was supported by grants and contracts from the National Institute of Allergy and Infectious Diseases, National Institutes of Health (AI-18357; AI-05074; AI-30189).

Literature Cited

1. Kochi, A. *Tubercle* **1991**, *72*, 1-6.
2. Horsburg, C. R.; Selik, R. M. *Am. Rev. Respir. Dis.* **1989**, *139*, 4-7.
3. Brennan, P. J. In *Microbial Lipids;* Leive, L.; Schlessinger, D., Eds.; Academic Press, London, 1988, Vol. I; pp 203-298.
4. Chan, J.; Fan, X.; Hunter, S. W.; Brennan, P. J.; Bloom, B. R. *Infect. Immun.* **1991**, *59*, 1755-1761.
5. Gaylord, H.; Brennan, P. J. *Ann. Rev. Microbiol.* **1987**, *41*, 645-675.
6. Hunter, S. W.; Murphy, R. C.; Clay, K.; Goren, M. B.; Brennan, P. J. *J. Biol. Chem.* **1983**, *258*, 10481- 10487.
7. Hunter, S. W.; Jardine, I.; Yanagihara, D. L.; Brennan, P. J. *Biochemistry* **1985**, *24*, 2798-2805.
8. Brennan, P. J.; Goren, M. B. *J. Biol. Chem.* **1979**, *254*, 4205-4211.
9. Saadat, S.; Ballou, C. E. *J. Biol. Chem.* **1983**, *258*, 1813-1818.
10. Besra, G. S.; Suzuki, R.; McNeil, M. R.; Khoo, K.-H.; Dell, A.; Brennan, P. J. Submitted for publication.
11. Schaefer, W. B. *Am. Rev. Respir. Dis.* **1967**, *92*, 85-93.
12. Hunter, S. W.; Fujiwara, T.; Murphy, R. C.; Brennan, P. J. *J. Biol. Chem.* **1984**, *259*, 9729-9734.
13. Hunter, S. W.; Barr, V. L.; McNeil, M.; Jardine, I.; Brennan, P. J. *Biochemistry* **1988**, *27*, 1549-1556.
14. McNeil, M.; Tsang, A. Y.; McClatchy, J. K.; Stewart, C.; Jardine, I.; Brennan, P. J. *J. Bacteriol.* **1987**, *169*, 3312-3320.
15. Daffe, M.; McNeil, M.; Brennan, P. J. *Biochemistry* **1991**, *30*, 378-388.
16. Besra, G. S.; Khoo, K.-H.; McNeil, M. R.; Dell, A.; Morris, H. R.; Brennan, P. J. Submitted for publication.
17. Camphausen, R. T.; McNeil, M.; Jardine, I.; Brennan, P. J. *J. Bacteriol.* **1987**, *169*, 5473-5480.
18. Besra, G. S.; McNeil, M. R.; Brennan, P. J. *Biochemistry* **1992**, *31*, 6504-6509.
19. Dobson, A.; Minnikin, D. E.; Minnikin, S. M.; Parlett, J. H.; Goodfellow, M.; Ridell, M.; Magnusson, M. In *Chemical Methods in Bacterial Systematics;* Goodfellow, M., and Minnikin, D. E., Eds.; Academic Press, London, **1985**, pp 237-265.
20. Minnikin, D. E.; Dobson, G.; Sesardic, D.; Ridell, M. *J. Gen. Microbiol.* **1985**, *131*, 1369-1374.

21. Ridell, M.; Wallerstrom, G.; Minnikin, D. E.; Bolten, R. C.; Magnusson, M. *Tubercle*, **1992**, *37*, 101-105.
22. Besra, G. S.; Bolton, R. C.; McNeil, M. R.; Ridell, M.; Simpson, K. E.; Glushka, J.; van Halbeek, H.; Brennan, P. J.; Minnikin, D. E. *Biochemistry* **1992**, 9832-9837.
23. Hunter, S. W.; Brennan, P. J. *J. Bacteriol.* **1981**, *147*, 728-735.
24. Mehra, V.; Brennan, P. J.; Rada, E.; Convit, Bloom, B. R. *Nature* **1984**, *308*, 194-196.
25. McNeil, M.; Gaylord, H.; Brennan, P. J. *Carbohydr. Res.* **1988**, *177*, 185-198.
26. Daffe, M.; Cho, S.-N.; Chatterjee, D.; Brennan, P. J. *J. Infect. Dis.* **1991**, *163*, 161-168.
27. Goren, M. B. *Am. Rev. Respir. Dis.* **1982**, *125*, 50-69.
28. McNeil, M.; Chatterjee, D.; Hunter, S. W.; Brennan, P. J. In *Methods in Enzymology;* Ginsburg, V., Ed.; Academic Press, San Diego, California, 1989; Vol. 179; pp. 215-242.
29. Bozic, C. M.; McNeil, M.; Chatterjee, D.; Jardine, I.; Brennan, P. J. *J. Biol. Chem.* **1988**, *263*, 14984- 14991.
30. Aspinall, G. O.; Khare, N. K.; Sood, R. K.; Chatterjee, D.; Rivoire, B.; Brennan, P. J. *Carbohydr. Res.* **1991**, *216*, 357-373.
31. Aspinall, G. O.; Crane, A. M.; Gammon, D. W.; Ibrahim, I. H.; Khare, N. K.; Chatterjee, D.; Rivoire, B.; Brennan, P. J. *Carbohydr. Res.* **1991**, *216*, 337-355.
32. Aspinall, G.O.; Gammon, D.W.: Sood, R. K.; Chatterjee, D.; Rivoire, B.; Brennan, P. J. *Carbohydr. Res.* **1992**, *237*, 57-77.
33. Takeo, K.; Aspinall, G. O.; Chatterjee, D.; Brennan, P. J. *Carbohydr. Res.* **1986**, *150*, 133-150.
34. Rivoire, B.; Ranchoff, B. J.; Chatterjee, D.; Gaylord, H.; Tsang, A. Y.; Kolk, A. H. J.; Aspinall, G. O.; Brennan, P. J. *Infect. Immun.* **1989**, *57*, 3147-3158.
35. Camphausen, R. T.; Jones, R. L.; Brennan, P. J. *Proc. Natl. Acad. Sci. USA* **1985**, *82*, 3068-3072.
36. Camphausen, R. T.; Jones, R. L.; Brennan, P. J. *J. Bacteriol.* **1986**, *168*, 660-667.
37. Tsang, A. Y.; Drupa, I.; Goldberg, M.; McClatchy, J. K.; Brennan, P. J. *Int. J. Syst. Bacteriol.* **1983**, *33*, 285-292.
38. Brennan, P. J. In *Microbiology 1984;* Leive, L.; Schlessinger, D., Eds.; American Society for Microbiology, Washington, D.C., **1984**; pp 366-375.
39. Lopez-Marin, L. M.; Laneelle, M.-A.; Prome, D.; Daffe, M.; Laneelle, G.; Prome, J.-C. *Biochemistry* **1991**, *30*, 10536-10542.
40. Besra, G. S.; Gurcha, S. S.; Hamid, M. E. A.; Goodfellow, M.; Minnikin, D. E.; Brennan, P. J. Unpublished results.
41. Jenkins, P. A.; Marks, J.; Schaefer, W. B. *Tubercle* **1972**, *53*, 118-127.
42. Riviere, M.; Puzo, G. *J. Biol. Chem.* **1991**, *266*, 9057-9063.
43. Riviere, M.; Puzo, G. *Biochemistry* **1992**, *31*, 3575-3580.
44. Besra, G. S.; McNeil, M. R.; Rivoire, B.; Khoo, K.-H.; Morris, H. R.; Dell, A.; Brennan, P. J. *Biochemistry* **1993**, *32*, 347-355.

45. Folch, J.; Lees, M.; Sloane-Stanley, G. H. *J. Biol. Chem.* **1957**, *226*, 497-509.
46. McNeil, M.; Tsang, A. Y.; Brennan, P. J. *J. Biol. Chem.* **1987**, *262*, 2630-2635.
47. Gerwig, G. J.; Kamerling, J. P.; Vliegenthart, J. F. G. *Carbohydr. Res.* **1978**, *62*, 349-357.
48. MacKenzie, S. L.; Tenaschuk, D. J. *J. Chromatogr.* **1975**, *111*, 413-415.
49. Hirschfield, G. R.; McNeil, M.; Brennan, P. J. *J. Bacteriol.* **1990**, *172*, 1005-1013.
50. Stellner, K.; Saito, H.; Hakomori, S.-I. *Arch. Biochem. Biophys.* **1973**, *155*, 464-472.
51. Hakomori, S. *J. Biochem.* **1964**, *55*, 205-208.
52. York, W. S.; Darvill, A. G.; McNeil, M.; Stevenson, J. T.; Albersheim, P. *Methods Enzymol.* **1986**, *118*, 3-40.
53. Prehm, P. *Carbohydr. Res.* **1980**, *78*, 372-374.
54. York, W. S.; McNeil, M.; Darvill, A. G.; Albersheim, P. *J. Bacteriol.* **1980**, *142*, 243-248.
55. Valent, B. S.; Darvill, A. G.; McNeil, M.; Robertson, B. K.; Albersheim, P. *Carbohydr. Res.* **1980**, *79*, 165-192.
56. Brennan, P. J.; Mayer, H.; Aspinall, G. O.; Nam Shin, J. E. *Eur. J. Biochem.* **1981**, *115*, 7-15.
57. Dell, A. *Methods Enzymol.* **1990**, *193*, 647-660.
58. Jardine, I.; Scanlan, G.; McNeil, M.; Brennan, P. J. *Anal. Chem.* **1989**, *61*, 416-422.
59. Khoo, K.-H.; Maizels, R. M.; Page, A. P.; Taylor, G. W.; Rendell, N. B.; Dell, A. *Glycobiology* **1991**, *1*, 163-171.
60. Dell, A.; Thomas-Oates, J. E.; Rogers, H. E.; Tiller, P. R. *Biochimie* **1988**, *70*, 1435-1444.
61. Besra, G. S.; McNeil, M.; Minnikin, D.; Portaels, F.; Ridell, M.; Brennan, P. J. *Biochemistry* **1991**, *30*, 7772-7777.
62. Chatterjee, D.; Bozic, C.; Aspinall, G. O.; Brennan, P. J. *J. Biol. Chem.* **1988**, *263*, 4092-4097.
63. Yoshimura, A. A.; Sato, K. I.; Singh, R. B.; Hashimoto, H. *Carbohydr. Res.* **1987**, *166*, 253-262.
64. Jansson, P. E.; Kenne, L.; Liedgren, H.; Lindberg, B.; Lonngren, J. *Chemical Commun.* **1976**, *8*, 1-75.
65. Lemieux, R. U.; Stevens, J. D. *Can. J. Chem.* **1966**, *44*, 249-262.
66. Bebault, G. M.; Duttan, G. G. S.; Funnell, N. A.; Mackie, K. L. *Carbohydr. Res.* **1978**, *63*, 183-192.
67. Dell, A.; Thomas-Oates, J. E. In *Analysis of Carbohydrates by GC and MS;* Biermann, C. J.; McGinnis, G. D., Eds.; CRC Press, Inc., Boca Raton, Florida, **1989**; pp 217-235.
68. Chatterjee, D.; Bozic, C. M.; Knisley, C.; Cho, S.-N.; Brennan, P. J. *Infect. Immun.* **1989**, *57*, 322-330.
69. Daffe, M.; Lacave, C.; Laneelle, M.-A.; Laneelle, G. *Eur. J. Biochem.* **1987**, *167*, 155-160.

70. Hunter, S. W.; Fujiwara, T.; Brennan, P. J. *J. Biol. Chem.* **1982**, *257*, 15072-15078.
71. Ville, C.; Gastambide-Odier, M. *Carbohydr. Res.* **1970**, *12*, 97-107.
72. Fournie, J. J., Riviere, M.; Puzo, G. *J. Biol. Chem.* **1987**, *262*, 3174-3179.
73. Riviere, M.; Fournie, J. J.; Puzo, G. *J. Biol. Chem.* **1987**, *262*, 14879-14884.
74. Daffe, M.; Varnerot, A.; Levy-Frebault, V. V. *J. Gen. Microbiol.* **1992**, *138*, 131-137.
75. Navalkar, R. G.; Wiegeshaus, E.; Kando, E.; Kim, H. K.; Smith, D. W. *J. Bacteriol.* **1965**, *90*, 262-265.
76. Sarda, P.; Gastambide-Odier, M. *Chem. Phys. Lipids* **1967**, *1*, 434-444.
77. Vercellone, A.; Riviere, M.; Fournie, J. J.; Puzo, G. *Chem. Phys. Lipids* **1988**, *48*, 129-134.
78. Marin, L. M. L.; Promé, D.; Laneelle, M. A.; Daffe, M.; Promé, J. C. *J. Am. Chem. Soc. Mass Spectrom.* **1992**, *3*, 556-561.

RECEIVED July 8, 1993

Author Index

Affiliation Index

Subject Index

Production: Donna Lucas
Indexing: Deborah H. Steiner
Acquisition: Rhonda Bitterli
Cover design: Amy Meyer Phifer

Printed and bound by Maple Press, York, PA